Into the Past

Into the Past

A memoir

Phillip Tobias

Foreword by Sydney Brenner

First published in 2005 by
Picador Africa
(an imprint of Pan Macmillan South Africa)
1st Floor, The Pavilion, Wanderers Office Park,
52 Corlett Drive, Illovo, Johannesburg, 2196
www.picadorafrica.com

and Wits University Press
1 Jan Smuts Avenue, Johannesburg, 2001
http://witspress.wits.ac.za

ISBN 1-7701001-5-6

Edited by Andrea Nattrass
Typeset in 11pt Minion by Abdul Amien
Cover design by Kevin Shenton of Triple M Design and Advertising
Cover photograph by Simon Hunt
Printed and bound in South Africa by Paarl Print

Contents

Acknowledgements *ix*
Foreword *xv*
Preface *1*

---

1 Dark days in Durban *3*
2 Benign banishment *8*
3 Formative years *14*
4 Into the arms of medicine *22*
5 Getting to grips with genetics *29*
6 The lure of Makapansgat *37*
7 Unravelling the past in Mwulu's Cave *45*
8 Hello to Sterkfontein *49*
9 Political awakening *55*
10 African eye-opener *63*
11 The happiest year of my life *70*
12 The Duckworth Laboratory *76*
13 French interludes *83*
14 Last months in Cambridge *92*
15 Arriving in the USA *98*
16 Los Angeles and the Grand Canyon *105*
17 Sojourn in St Louis *111*
18 Human genetics in Ann Arbor *118*
19 The whirlwind tour continues *124*
20 Hot and humid days in Washington *133*
21 More East Coast travels *141*
22 New York City and Cold Spring Harbor *149*
23 The Tonga people and the Kariba Dam *157*

24 Growing up or growing down? *165*
25 A reluctant professor *174*
26 Apartheid in the university arena *184*
27 Academic boycotts and freedoms *193*
28 Joyous variety *202*
29 My most unforgettable character *211*
30 The Darwin of the twentieth century *221*
31 Travelling fossils *228*
32 To whom do the fossil hominids belong? *237*

Notes *251*
Select bibliography *253*
Index *263*

Photographs fall between pages 144 and 145.

For Joe and Isa, Heather, Irene, Samuel and Martha,
to each of whom I owe so much.

With heartfelt thanks to my beloved parents,
sister and stepfather/godfather.

In grateful memory of Raymond Dart, Joseph Gillman,
Robert Broom, Louis and Mary Leakey,
Wilfrid LeGros Clark and Theodosius Dobzhansky.

To my 10 000 students who helped to transfigure
my dreams into reality and, after all, kept me
tolerably young.

Acknowledgements

This is a book mainly about the first half of my life. The thanks I owe are to those who helped me live those 40 years and to those who assisted me in preparing this first autobiographical volume. It is the mark of time that in the former group, many have died who were once so meaningful to me.

I think with nostalgia and deep appreciation of schoolteachers in Durban and Bloemfontein. Those at President Brand School and St Andrews School in Bloemfontein guided me in the years leading to puberty. Among many others, there were 'Dassie' Britten, who helped me to master my fear of water and introduced me to an understanding of the human body's immensely complicated yet superbly integrated pattern; Frank 'Oubaas' Storey who at 'Saints' ran a tight ship; John Parry for whom a mastery of the King's English was an all-encompassing goal in life; and Roger C. Bone who perhaps never realised how much of my life he catalysed when I held my first editorship under his encouraging eye. At Durban High School, I shall be forever indebted to 'Nev' Nuttall for ushering me into the portals of English literature; 'Charlie' Evans who fired me with a lifelong love of history; Neville Bowden who taught us how logical the Latin language was and enlivened it with flashes of ribald humour; 'Little Van' (Van Heerden) and, especially, 'Big Van' (Van Reenen) who brought us to see the flow and lilt of the Afrikaans tongue; and 'Jimmie' Black who, as headmaster, was sufficiently broadminded to appoint as a school prefect someone who was not an outstanding sportsman but in whom he must have seen other signs of leadership.

A loving and tolerant family complemented my 'school families'. To my dear sister Valerie and loving parents Joe Tobias and Fanny Rosendorff, I must add that most understanding of men, Bert Norden, my godfather who later became my stepfather.

I extend my gratitude to those who helped to shape me at Wits University. It is a veritable Senatus Academicus of people such as 'Pop' van der Horst, Henry Stephen, John Phillips, Brian Farrell, and the 'First Eleven' in the Anatomy Department: Raymond Dart, Joe Gillman, Lawrie Wells, 'Sandy' Galloway, Christine Gilbert, George Oettle, Mike Wright, Phyllis Knocker, Joel Mandelstam, 'Toby' Arnold and 'Jos' Lannon, and the 'Second Eleven' comprising Adèle Trope, Sarah Klempman, Doris Bronks, Sydney Brenner, Barbara Allison, Pris Kincaid-Smith, Marianne Cassirer, Julien Hoffman, Percy Barkham, Dan Goldstein and Jan Toerien – many of this second team are my near-contemporaries and still alive. Later respected stewards of the Anatomy Department were 'Benjy' Rawdon, Maciej Henneberg, Noel Cameron, Beverley Kramer and John Maina.

My physiology teachers were A. Dighton Stammers, Sonia Highman, Ralph Bernstein, Cecil Percy Luck, 'Benny' Bloomberg, Sydney ('Larkie') Cohen and Benny Kaminer, while we were introduced to pathology by 'Archie' Strachan, 'Chrissie' Chatgidakis and 'Bunny' Becker. The realms of pharmacology were displayed to us by Professor J. Watt and 'Flossie' Stephen-Lewis. After I had survived the preclinical and paraclinical years, it was the clinicians into whose spheres of influence I tiptoed: Guy Elliot, 'Jock' Gear, David Ordman, 'Cocky' Underwood, Tony Leonsinis, Lee McGregor, 'Jock' Wolfowitz, Jack Allan, Steve 'Ossie' Heyns and 'Fossie' Daubenton.

To all of these teachers – and to those whose names I have inadvertently or absent-mindedly omitted – my heartfelt thanks are due.

Of course it was not only my mentors and friends at the Wits Medical School to whom I extend my thanks. There were those from other institutions and from far afield including: Andrew Abbie, Emiliano Aguirre, Leslie Aiello, Peter Andrews, Camille Arambourg, Robert Ardrey, Kader Asmal, Francisco Ayala, Hisao Baba, Ali Bacher, Francois Balsan, Nigel Barnicot, D. Benghu, Alan Bilsborough, Joe Birdsell, 'Teddy' Boné, Gerhard von Bonin, Terry Borain, G.R. 'Boz' Bozzoli, 'Bob' Brain, Gunter Brauer, Robert Broom, Nils Burwitz, Karl Butzer, Bernard Campbell, Jean Chavaillon, Shiv Chopra, Desmond Clark, Wilfrid LeGros Clark, 'Monty' Cobb, 'Bert' Cohen, Juan Comas, Glenn Conroy, Basil Cooke, Carleton Coon, Yves Coppens, Pamela Tilden Davies, Michael Day, Pamela de Beer-Kaufman, Henry and Marie-Antoinette de Lumley, Hertha de Villiers, John de Vos, Dialo Diop, Theodosius Dobzhansky, 'Sonny' du Plessis, B. Endo, 'Griff' Ewer, Denise Ferembach, Leo Gabunia, George Gaylord Simpson, E. Genet-Varcin, Santiago Genoves,

Asok Ghosh, José Gibert, Barbara Goldberg, Nadine Gordimer, I. 'Okkie' Gordon, 'Bill' Greulich, Georg Haas, Kazuro Hanihara, H. Hemmer, Ruthie Hoffman, Ralph Holloway, Barbara Honeyman-Heath, F. Clark Howell, 'Bill' Howells, John Huizinga, 'Solly' Hurwitz, Andrew Huxley, Kinji Imanishi, H. Ishida, Junichiro Itani, Teuku Jacob, Trefor Jenkins, Don Johanson, Frank Johnston, Clifford Jolly, Kenneth Kennedy, G.F. Khroustov, 'Jules' Kieser, 'Bill' Kimbel, Lester King, Ralph von Koenigswald, V. Kochetkova, P.C. Koller, Grover Krantz, Michel Kretzoi, Gottfried Kurth, Lorraine Lake, Louis S.B. Leakey, Mary Leakey, Meave Leakey, Richard Leakey, Harding Le Riche, David Lewis-Williams, Graham and Barbara Lindop, Frank Livingston, C. Owen Lovejoy, C. van Riet Lowe, Bridget Mabandla, I.D. MacCrone, Neil Mackintosh, Vincent Maglio, Sajiro Makino, 'Berry' Malan, Alan Mann, Z.K. Matthews, Robert Matthey, Ernst Mayr, Peter Maytom, P.V. Mbatha, 'Ted' McCown, Henry McHenry, Sarah G. Millin, Yuji Mizoguchi, Theya Molleson, Giuseppe Montalenti, N.G. Moodley, Geoffrey Morant, Alan Morris, Sydney Motlokoane, A.E. Mourant, Alan Muir, John Napier, Hillel Nathan, James Neel, Karel Nel, Ben Ngubane, Kenneth Oakley, 'Pro' Obel, Georges Olivier, Keiichi Omoto, Tim Partridge, Bradley Patten, Roy Peterson, Nicola Petit-Maire, David Pilbeam, Jean Piveteau, Rosemary Powers, Phillip Rightmire, John Robinson, George Romanes, Avraham Ronen, Ann Rowe, Wu Rukang, Karl Saller, Vincent Sarich, S. Sartono, Gerrit W.H. Schepers, Friedemann Schrenk, Horst Seidler, Sergio Sergi, Geoff Sharman, Elwyn Simons, John Skinner, Pamela Solarsh, 'Kris' Somers, Dolly Spence, Geoff Sperber, Moishe Stekelis, Dale Stewart, 'Bill' Straus, Daris Swindler, Helen Suzman, Jiro Tanaka, Ian Tattersall, Teilhard de Chardin, Robert J. Terry, Andor Thoma, Miriam Tildesley, John Tinker, Jack Trevor, Mildred Trotter, Daniel Turbon, Russell Tuttle, Henri Vallois, Laszlo Vértès, Emman Vlcek, John Vogel, Elisabeth Vrba, Conrad Waddington, Alan Walker, 'Sherry' Washburn, Joe Weiner, Stella Welsh, 'Vic' Wessels, Tim White, Napoleon Wolanski, Brian Wolfowitz, Milford Wolpoff, Bernard Wood, V.P. Yakimov, Adrienne Zihlman, Eduard van Zinderen Bakker, Solly Zuckerman and Saul Zwi.

Individually these scholars helped to mould me, not only by their friendship, although that was always very important, but through their several insights, varying approaches and many windows their works opened for me.

Over the years a succession of research assistants greatly helped in the development and unravelling of my research activities. They included Adèle Munro, Theo Badenhuizen, Jenni Soussi (later Soussi-Tsafrir), Brian Maguire, Edith Hibbett, Noreen Gruskin, Carole Orkin, Val Strong, Heather White, Kay Copley and more besides. What I have achieved I could not have done without their devoted help over the years.

Others who served the Anatomy Department and me over many years deserve my heartfelt appreciation. The roles fulfilled by Alun Hughes, Roland Klomfass, 'Kaufie' Kaufman, Peter Faugust, Syd Dry, Norman Harington, John Bunning, Les Hastings, Ronnie Herman and Brian Hume, cannot be overemphasised. The secretaries have been sempiternally dependable: often they told me how much they enjoyed working for the department and me. Among them – even though I know it is invidious to single out some – Bernice Wilson, Hazel Ball, Paddy Driman, Sheila Boles, Marlene Talbot, Anya the sky-diver, June Asch, Zubeida Jeewa, Joan Ward, Maureen Schneiderman, Eileen Judd and Heather White stand out in my memory.

Ann Andrew, Jack Allan, Mary Veenstra, Ron Clarke, Gabriel Macho and others helped to keep the department, the research unit and me functioning efficiently.

On the domestic front, over the last 45 years, Gezani Samuel Shirinda, Martha Dlamini and Irene Jones have at various times kept the home-fires burning. Their loyalty has been equalled only by their friendship. Sydney Motlokoane cared for me, especially in the later years, in my medical school office. To these 'personal assistants' must be added my general practitioner, Dr Joe Teeger, sublimely unusual in his dedicated and unselfish care for his patients. I like to say of him that he has kept me breathing for half a century or more! What good fortune I have been blessed with to have these gentle people watching over me.

The help, co-operation and friendship I have received from scores of research students – including my B.Sc., B.Sc. Hons, M.Sc., doctoral and post-doctoral students from South Africa and other parts of the world – over the last half-century have been one of the pillars of myself, my department and my researches. Although they are too numerous to name here, my gratitude to them is overwhelming.

The preparation of this book was first proposed to me by Gerald de Villiers of Ravan Press and Pat Tucker of Wits University Press some six years ago. We signed a contract on 4 February 1999 and then several years elapsed before my scale of activities and my frame of mind permitted me to start writing – on 11 January 2004. Long before the end of that year it had become evident that more than one book was beginning to take shape. My publishers were very tolerant and patient with me despite the elapsing barren years. Now, as the goal is in sight (at least in respect of the first volume), it is an enormous pleasure for me to put on record my indebtedness to a wonderful team of people: for Pan MacMillan they have been Gerald de Villiers, Dusanka Stojakovic and Terry Morris, and for Wits University Press, Pat Tucker at an early stage and Veronica Klipp. It has been a great pleasure to work with such enthusiastic, creative and positive-minded people.

Andrea Nattrass has been undaunted in rendering my over-long manuscript into a terse and readable book. She has been thoughtful, patient and forbearing in the face of my literary and grammatical foibles. Although she is in Pietermaritzburg and I am in Johannesburg, she allowed no considerations of geography or family commitments to interfere with, or retard, the smooth and swift completion of the work on my manuscript. Such errors as must inevitably have crept in are, however, to be laid at my door, not hers, and I take full responsibility for them.

Special gratitude is due to Peter Faugust in the School of Anatomical Sciences, Terry Borain in the illustration unit and Graeme Williams.

Several people have served as informal consultants to me over the years, or specifically in the preparation of this book: Bruce Murray, Noel Garson, Trefor Jenkins, Georgio Manzi, Teuku Jacob, Archbishop Denis Hurley, Father Franklin Ewing, Francis Thackeray and Ron Clarke. Thanks to all of you and others I should have mentioned!

Finally, words fail me when I wish to sing the praises of Heather White, my personal assistant for the past twenty years. The whole stage-managing of this project has revolved around her uncomplaining and ever-willing hands, shoulders, eyes, ears and brain. Thank you Heatherbelle!

Foreword

Many years ago, when we were students in the Old Medical School at Wits, I very much envied Phillip Tobias for his facility with words and the ease that he displayed in assembling them into written and spoken language. He describes in this memoir how he came to write it. It seems that although he had some doubts about the general nature of the project, when he decided to do it, he simply sat down and wrote it. Not so with the writer of this Foreword who has always been pen-tied, never satisfied with the opening sentence, and who finds every excuse not to surmount the difficulty. It seems most inappropriate to have this stumbling introduction to a book that speaks lucidly and inspiringly for itself.

I have heard it said that good book reviewers never read the book being reviewed, because that might distract them from the main task of telling the world about themselves. Writers of forewords should keep this in mind, but what is more important is that they should not tell the reader too much about what follows, even though they have read it. No excuse should be provided to readers not to read the book itself. However, a few things should be said. Unlike many of his contemporaries who left South Africa in the 1950s, Phillip stayed on and committed himself to maintaining high standards of scholarship and personal integrity during the difficult years. His memoir not only records his personal journey in life, science and education, but also the passage of our country, South Africa, now so miraculously changed and embarking on new paths of social discovery.

The young today are not interested in history; the past is constituted for them only by the last two or three years. Those of us who are biologists know that everything is rooted in the past and that one way of understanding what we are is to know where we came from.

As one grows older, one comes to value memoirs, reminiscences and diaries for the personal view that illuminates the past and confers an individual humanity on it.

Sydney Brenner
Nobel Laureate

Preface

'Don't you think it's time to write your autobiography?' a sports journalist asked boxer Chris Eubank. The British Middle-Weight World Champion looked puzzled. 'About what?' he asked.

This is the kind of question I cannot escape as I embark on writing my memoirs. It is six years since I signed the contract to prepare an autobiography.

I had misgivings then, just as I had doubts the last time such a contract decorated my desk fifteen years ago.

Nevertheless, not long after my 74th birthday, I sat under the palm trees looking out over the bay of Palma de Mallorca. It was a mild, moist, Mediterranean evening. I started to write:

> Thursday evening, 10th February 2000
>
> At the Mirador Hotel, Palma de Mallorca. This clean, virginally white, first page of a sketchbook lies before me enticingly, seductively. Nils Burwitz gave it to me this morning, urging me to use it to start my autobiography . . . I think of another beautiful, Florentine book, with hand-made paper, that Jacopo Moggi Cecchi gave me as a gift in April 1996, for the same purpose. The pages of that book remain unsullied. Why do I have such hesitation, such reluctance, to start writing about my life and times? I have always enjoyed writing and have found it easy, undemanding, to write about others – Raymond Dart, Robert Broom, Ralph von Koenigswald, Louis and Mary Leakey, Sydney Brenner, John Robinson, Frank Spencer, Lawrence Wells and many others. Why then is it so hard for me to step out of myself, stand back a little, and limn a picture of me?

I sit back and contemplate the situation. Is it not a monstrous conceit to think that others may want to read about my life? How good a portrait does an autobiography paint of a person? Its inevitable selectivity frightens me. Keep it short, say my publishers, perhaps unaware of how difficult this has been for me all my life. What of the pieces and people I leave out? Isn't there a danger that a personal chronicle may become a litany of self-glorification? I shudder at the thought: it goes against the grain. It should surely be the story of a person warts and all. Who am I to describe my warts, blemishes and imperfections? The narrative must obviously be subjective – and perhaps this is the sticking place to which I have been loath to screw my courage. As a practising scientist for more than half a century, I have been trained, and I have tried, to be objective. Then again, an autobiography – even of a scientist – is not a piece of scientific writing. So the argument goes back and forth in my teeming brain and I find myself once again unable to continue to put pen to paper.

FOUR YEARS LATER in my Johannesburg flat, near the famous Wanderers cricket ground, I reopen the sketchbook. Then I open my mind and start typing on my Olympia electric typewriter – which elicits more from me than my computer ever could. Whatever my doubts and reservations, I have decided that there can be no more waiting. It is time for me to tackle the daunting task of writing my autobiography – it's now or never!

1 Dark days in Durban

I was born by the sea, the warm Indian Ocean caressing the coast of what is now KwaZulu-Natal. Although it has been my destiny to spend my adult life in Johannesburg, nearly 2 000 metres above sea level, I have never lost my love for the ocean. Not only do I like to spend time on the long, crenulated coastline of South Africa, but visits to islands always give me great joy too. The only time that I am able to relax is when I am gazing at the breakers, sniffing sea air, or immersing myself in salt water.

When I was a schoolboy, there were no shark-nets along the Natal coast and the thought of being carried off, as swimmers were now and again, gave an added thrill to the swimming experience.

My family fell on hard times, both economic and social, during those early Durban days. My father, Joseph Newman Tobias, was born in Portsea, England, in 1884. During and shortly after the South African War of 1899 to 1902 (formerly known as the Anglo-Boer War), the family came out to what was then the Colony of Natal. My grandfather, Phillip Tobias, opened a toyshop called Tobias and Sons Bazaar in West Street, the main shopping thoroughfare of Durban.

Two of my uncles returned to England and my father inherited the toyshop. As a child I used to go down to father's shop and ride on the scooters, motor cars and tricycles that were for sale. In those early years, we had a double-storey house in Hayle Avenue up on the Berea. My mother had a small upright piano and she used to say that she played out all her emotions on it. She was a qualified piano teacher and had acquired a diploma from the College of Preceptors. In her younger days, she gave piano lessons in Cape Town. Later, when the family's fortunes failed, she resumed teaching in Bloemfontein, near her birthplace of Edenburg, in the Free State (formerly the Orange Free State). Her maiden name was Fanny Rosendorff and she was born on 12 August 1891, the oldest of seven children. Her father, Martin Rosendorff, had been born in Berlin, Germany, in 1855 and came to South Africa in 1877.

The two families, Tobias from England and Rosendorff from Germany, were united when my father and mother married in Bloemfontein in 1919.

SOME OF MY earliest memories are of friction between my parents. The birth of my sister Valerie Pearl Tobias on New Year's day in 1921 and of myself on 14 October 1925 were not sufficient to bond my parents. I think that there was a clash of personalities and there were always financial difficulties. The toyshop had flourished in earlier years, but when the wholesale bazaars opened up in Durban, the competition proved too great and my father was reduced to bankruptcy. The toyshop closed down, the house was sold, and the four of us moved to the Ivanhoe, a small residential hotel or boarding-house, as it was called in those days.

It was a gloomy period. My parents and I shared one room. Although a free-standing screen was placed around my bed, I could not escape overhearing their nightly arguments and wrangling. My sister Val was enrolled in the Convent of the Holy Family close by and she excelled in French. She was vivacious and dynamic, a talented tap-dancer and singer, and as a child often performed on the local stage.

Apart from the satisfaction my father derived from my sister's and my successes at our respective schools, his only consolations seemed to be playing snooker and billiards with one or two close friends, and attending cricket matches at the famous Kingsmead Cricket Ground in Durban. He found it difficult to obtain a new job and wondered if Johannesburg would be more promising. The plan was that he would go up to the 'Golden City' and when he had found something suitable, the rest of the family would follow him. The outcome was catastrophic. The positions that he found, as an estate agent with his brother-in-law at one time, and running a pleasant little tearoom in downtown Johannesburg at another, and the accommodation that he offered my mother in a boarding-house, were unacceptable to her. She returned to Durban in tears and it was clear that the marriage was irrevocably harmed.

As a precocious and sensitive child, I was well aware of these painful developments, at least from the age of five or six. Bouts of depression and night terrors punctuated my early years. Nevertheless, with something approaching religious ardour, I idealised the concept of a happy married home and the contented life of a joyous and blessed family. Although I never personally enjoyed such halcyon times, a strong streak of positive thinking did not allow me to let go of this elusive vision. I clung to a verse that 'Jean' wrote in my autograph album in January 1937 when I was eleven years old:

> Don't look for the flaws as you go through life,
> And even if you find them,
> 'Twere wise and kind to be somewhat blind
> And look for the virtues behind them.

I could not take sides in the parental clash, and I could not determine who was right and who was wrong. I loved both my mother and father. From my 13th year I stayed part of each year with my father and part with my mother. During my high school days in Durban, I spent the school terms with my mother and stepfather in Westville, outside Durban, and the school holidays with my father in Johannesburg. When I started studying at the University of the Witwatersrand (Wits), I stayed with my father in Johannesburg and spent the university vacations in Durban. It was a reasonable compromise and life went on in this way until my father died in 1963 (aged 78) and my mother in 1988 (aged 97).

AFTER FATHER WENT to live in Johannesburg, my mother, sister and I transferred by rickshaw to another residential hotel. I have never forgotten that ignominious ride. Mother's tears cast my sister and me into our own moist-eyed gloom. We settled at the Coogee Beach Hotel in Tyzack Street, close to Durban's South Beach. I had to take two tram rides each way – one to the post office, and the second one up to Gordon Road – to get to Durban Preparatory and back each day. I loved writing and it was in my first and second years at this school (when I was seven and eight) that I published my earliest fragments in *Prep's Own Paper*. I also took up long-distance running, for which my slight frame was well suited, and enjoyed the school and inter-school competitions that were held.

We were living from hand to mouth and the parents of some of my school friends helped me out with school uniforms, books and even the season ticket for the daily tram rides. I remember feeling ashamed at having to receive this kind of charity, but my sense of shame only impelled me to greater efforts at school.

When I was nine, my third year at Durban Prep was seriously interrupted by an unusual accident. My mother sent me with a message to a friend at the Cecil, another beachfront hotel. I had to go up in the lift for several floors. It was one of those old-fashioned elevators in which the 'doors' were a pair of metal grilles. The floor was depressed when a passenger stepped on it and then the lift would not respond to a call from another floor. It may be that my light weight was scarcely enough for this under-floor mechanism to work properly. Whatever the reason, the lift went up for a floor or two and then stopped short of the next floor by about a metre. It would not go up or down and I felt a rising sense of panic. I climbed up to the brink of the landing above to try to open the latticed screen door. The inevitable happened: the moment my weight was released from the lift floor as I hoisted myself up, the elevator responded to a call from below. It descended, trapping me between the ceiling of the lift and the edge of the floor landing. My small body, mainly my right thigh, my right Achilles tendon and the back of my head bore the whole weight of the lift. Fortunately, I did not lose consciousness and was able to scream 'Help!' for all I was worth. It seemed an age before help arrived. The floor of the elevator was by now near the landing of the floor below. Somehow a man in the hotel managed to wedge his way into the lift from the next floor down. Then the lift was directed upwards, thus releasing me. He lifted me – a mass of blood and shock – down to the floor. Immediately I sensed that my right leg was broken, telling the man, 'I can't stand; my leg is broken'. He carried me across, blood pouring from an injury to the back of my head, to the hotel where we were staying.

The upshot was 93 days spent in Addington Hospital, with a simple fracture of the upper third of the right femur (I learned the diagnosis off by heart!), together with a partial severance of my right Achilles tendon and superficial scalp wounds. During my three months in the children's ward, I learned to respect and love the nurses, especially a night-nurse whose name I remember as Nurse Dawning. My right lower limb was encased in a full-length plaster cast and was up in the air on traction. My foot, far away from the reach of my hands, used to itch fearfully and the skin flaked. Nurse Dawning's long nails used to scratch that faraway foot and I could not sleep until she had done her rounds and brought relief to my isolated extremity.

Most of the boys in my ward were older than I was. From them I learned a fair amount about sex and some especially salacious jokes and foul words – or so they seemed to a cosseted youth of nine who was eager to learn about everything.

My father travelled down from Johannesburg to visit me in Addington and spent a few days in Durban. However, I became aware that my mother was seeing a great deal of my godfather, a man named Albert Louis Norden. She was earning a little money as a counter-hand at a shop in West Street and 'Uncle Bert' used to bring her to visit me in hospital. With a sense of delicacy, he always waited in his car while she spent the permitted half-hour at my bedside. He was a very fine man, a chartered accountant of great integrity and a descendant of the 1820 Settlers.

I emerged from hospital to find the whole of Durban *en fête* to celebrate the Silver Jubilee of King George V, the British monarch who then was also the king of South Africa. Bert invited my mother and me to stay at his country estate in Westville. There I grew strong and well and put on some of the weight I had lost. By the time I had convalesced, I had missed six months of schooling, but managed to make up the work when I returned to Durban Prep in the second half of the year.

By 1935 my parents' divorce was looming. To give my mother and father some space for the negotiations, my mother's family in Bloemfontein kindly agreed to take responsibility for Val and me. Thus it was that my first Durban phase ended on a sad note as we embarked on the overnight train journey from the lush, tropical ambience of the coast to the parched plateau of the interior.

2 Benign banishment

It was a mild culture shock to move from the very English city of Durban to the predominantly Afrikaans-speaking and -thinking Bloemfontein.

The terms 'English' and 'Afrikaans' essentially applied to the white or 'European' populations. The black African peoples of KwaZulu-Natal were in the main Zulu-speaking and those of the Free State were Sotho-speaking. Despite the fact that these two sectors of the populace were by far the majority in each of those two provinces, as indeed black Africans were in the Union of South Africa as a whole, it was as if they were on another planet. Socially, culturally, linguistically, politically and economically they were not treated as citizens of the country, not part of 'die Volk' (the people) of South Africa. Black Africans lived separately in 'locations' outside the cities and towns. They did not have the vote or even any say in determining their own affairs. Schools were segregated: 'white' schools excluded pupils of colour and 'black' schools were totally 'black'. Cinemas, theatres, concert halls, restaurants, hotels, universities and technical colleges, central, provincial and local government bodies were all strictly segregated. Public transport, such as buses, trams and trains, were for whites except for a few seats upstairs at the back of double-decker vehicles.

All of this was taken for granted by most people in the white community. At schools it was scarcely mentioned, if at all. Many textbooks, especially history books, perpetuated myths about the 'non-Europeans' and were peppered with pejorative descriptive terms.

The time I am speaking of in South Africa was the 1930s, well before the accession to power of a National Party government dedicated to the entrenchment of apartheid. The Afrikaans word 'apartheid' had not yet been assimilated into English dictionaries. Indeed, in the 1930s 'apartheid' was in use only in some academic and political circles. It was in the electioneering campaign that preceded the 1948 general election that it was first widely used.

In 1936, Bloemfontein was a quiet, pleasant and sleepy town, a far cry from the vibrancy of Africa's busiest port city of Durban and the wealth and economic hub that was Johannesburg. There was a restful atmosphere among a population given to concerts at home; picnicking on the Modder River, Salt Pan or at Bainsvlei; and local productions by repertory theatre, staged especially by Dolfanna Brown, wife of the dental specialist, Lester Brown. I played parts in two of her productions, *All God's Chillun got Wings* and *Snow White* (in which, as Prince Charming, I fell madly in love with Snow White, played by the producer's lovely daughter, Rochelle – who was later to become a fine flautist).

Val and I were entrusted to the care of my mother's youngest sister Dora, who was married to Hyam Posner. There were two Posner boys: Harry was eight weeks my senior, and Walter was two years younger. Their sister, June Elfrieda, was a talented and lovely little girl whom I idolised.

I went to the same school as the boys, the co-educational President Brand School. Not only did I encounter for the first time boys and girls in the same classes, but it was a dual-medium school: some classes were conducted in English and others, in parallel, in Afrikaans.

Nature study was one of our subjects and I found myself drawn to this first introduction to the world of 'living things'. Later I learned that this was the title of J.W.N. Sullivan's little book in Basic English, which explained even concepts as complex as evolution in simple, understandable terms.

For my next two years in Bloemfontein, the Posner boys and I attended the English-medium, Anglican St Andrew's School. One of the fine teachers at this school was Roger C. Bone, fresh from Cambridge. With his encouragement, we started a class newspaper, *Form Four Fortnightly*. I was the editor and wrote much of the 'copy', while Mr Bone typed it out on his portable, manual typewriter. It sold for a penny a time and was on sale every second Monday morning. *Form Four Fortnightly* provided my first editorial experience – when I was just twelve years old.

Little did I know then that I would spend a large part of my life writing, editing and publishing books and articles. In addition, many years later, I found myself checking manuscripts and draft theses of students and lecturers working under my supervision. There were the usual errors of logic, woolly definitions, unfamiliarity with relevant publications, poorly prepared or ill-defined research methods, inconsistencies of terminology, fallacious reasoning and muddled presentation. Those, of course, I expected. However, no fewer than 80 per cent of the mistakes were errors of English grammar, usage, idiom, choice of words, spelling, punctuation and the use of gobbledegook. I used up innumerable red pens in the correcting of such solecisms. My remarks about dangling participles, offensively split infinitives, ghastly errors of concord, malapropisms and other barbarisms, became notorious among my research students. Yet, although they might be the butt of furtive sniggers or fits of spleen, the message did get across – that there can be no clarity or logic of thought without correctness and lucidity of expression. Recently, a former American Ph.D. student, who had been through the mill with me, gratefully informed me that he was now returning his own students' manuscripts with similar scathing comments about the murdering of the Queen's English writ large between the lines and in the margins.

Not only students find it hard to express themselves correctly and elegantly in English. A practising scientist, in order to keep abreast of the subject, has to read scores of published articles each year. For someone like myself who has a love of English this task can be most vexing, for the standard of writing in the average medical or scientific journal is deplorable.

So bad was the abuse of the English language, although Wits was officially an English-medium university, that I was persuaded (somewhat against my better judgement) to present a series of talks for the Anatomy Department on 'Common Errors of English Usage'. These were intended for my own medical science and Honours students who grappled with the preparation of project reports. Soon, however, writers of theses and dissertations, university staff members whose home language was not English, and then even members of the public, joined the growing audiences. It started as a series of two or three lectures and grew over the years to six to eight lectures, which ranged from a potted history of the English language to Cockney rhyming slang and word games. What fun I had preparing and delivering these talks.

OUR WOLF-CUB TROOP'S akela (leader) in Bloemfontein in 1937 was 'Dassie' Britten. Apart from introducing us to the arcane world of junior scouts, he was the teacher of hygiene and nature study. In this course I became acquainted with the bones of the human skeleton and with other bits and pieces of human anatomy. The subject enthralled me and I still have some of my drawings of the eye, tooth, circulation of the blood, and the skeleton from that time.

When I look back to find the roots of my subsequent anatomical interests, my three years of nature study in Bloemfontein must be included amongst the various influences.

At first Val had stayed at the Posners' home and then she was sent to a famous girls' school, Eunice Institute, in the city. She was an enthusiastic and successful sportswoman and very popular in her early teens. I used to bicycle down to visit her at Eunice once a week. That is how I came to have so poignant a recollection of the onset of her diabetes. She would telephone me from school and ask me to bring her a bottle of fruit-juice. Then she would ring and ask me to bring two bottles each week. Her polydipsia grew and mid-week visits became necessary. Soon after that, her diabetes was diagnosed. She was sixteen and sadly was dead by the age of twenty-one. The treatment available in those days made her illness difficult to control, and there were also complicating psychological factors. She died in a nursing home in Port Shepstone on the Natal south coast on 4 May 1942. Indirectly, Val's illness and premature death played an extraordinary part in my life.

At that time I was sixteen years old and in my matric year at Durban High School. After Val's funeral, I recall talking to our family doctor. My maternal grandmother, Frieda Rosendorff, had died as a diabetic in 1928 when she was fifty-seven. Now Val had died of diabetes at twenty-one. If there was a genetic tendency for the disease it had manifestly come from grandmother through mother to my sister Val. Yet my mother was not a diabetic. I asked our family doctor how the diabetes gene or genes could have come through my mother, without her showing any sign of diabetes. He replied that he did not know. When I asked who in South Africa was versed in medical genetics, he said that, as far as he knew, there was no one. There and then I secretly resolved to become the first medical geneticist in South Africa. It was something that came to pass in the 1950s after I had completed my Ph.D. on cytogenetics (the study of chromosomes) and had studied heredity counselling under the leading human geneticist, James V. Neel, at the University of Michigan, Ann Arbor (see Chapter 18). This career had all started with my sorrowful puzzlement over Val's death. So often in life I have found that I have made major decisions based on illogical circumstances and even emotional reasons.

BACK IN BLOEMFONTEIN in 1937, Hyam and Dora Posner went away on a pre-war trip overseas. My mother came to Bloemfontein to look after the Posners' house and the children. She gave us piano lessons on Dora's upright piano. Cousin Harry Posner and I learned to play a duet, 'The Bluebells of Scotland'. I became aware that family gatherings of the Rosendorffs were taking place in the Posners' dining room. All the brothers were there – Max, Siegfried, Herman and Karlie. My mother confessed tearfully to me that they were negotiating about her future. My father was digging his heels in on some points of the divorce, probably on the question of custody of Val and me. Then, one day, mother asked me to come for a walk with her. I had just turned eleven and I knew what was coming, but didn't want to hear it. She told me that she and my father were divorcing and that she was going to marry Bert Norden. My dreams of a perfect family were shattered, my worst fears realised. I burst into tears and ran away from her. I locked my bedroom door and threw myself on my bed sobbing uncontrollably for hours.

As my Bloemfontein interlude was drawing to a close, the skies were brightened by the arrival of the English cricket team, known as the MCC team after the Marylebone Cricket Club. From my autograph album I see that they played the Orange Free State in November 1938. It was exciting for us thirteen-year-olds to go down to the Ramblers to watch the match, which the visitors won by an innings and 24 runs.

At the end of 1938, as ominous war clouds were growing over Europe, my junior schooling ended at St Andrew's. Although the principal, Frank Storey, wrote to my mother strongly urging that I stay on at Saint's senior school, the family had arranged for me to enter Durban High. With much loving gratitude to the Posners and all my cousins, as well as to St Andrew's for two wonderful years, I responded to the lure of the coast and found myself on the overnight train back to Durban.

There, early in 1939, I was to watch the MCC team play in the 'Timeless Test' at Kingsmead. This test match has been described as the longest and most remarkable test match ever played in the history of international cricket. It came to be known as the Timeless Test because it was agreed to play to a finish irrespective of the number of days taken. This amazing game, which I went down to Kingsmead to watch on every other day, occupied twelve days, including nine of actual play, one being washed out by rain. By the end of that time, there was still no result and the historical match had to be abandoned to enable the English cricketers to catch the train to Cape Town and then the ship back to England.

3 Formative years

Durban High School was a fine institution with an outstanding reputation for scholastic and sporting achievements. It was especially well known in sporting circles for its contributions to cricket, a game I have always followed with keen interest. Two of the most famous cricketers to come out of the school were the distinguished batsmen Barry Richards and Lee Irvine, but there were many others in all departments of the game.

On the side of scholastic accomplishments, there was the great writer and poet Roy Campbell, whose 'The Flaming Terrapin' (1924) led to his immediate recognition as the first (white) South African poet of international stature.

In the scientific field, the school's most considerable product was undoubtedly Aaron Klug who was dux of the school in 1941, one year ahead of me. His early studies at the Universities of the Witwatersrand and Cape Town led to his later distinguished researches in England on the crystallographic analysis of biological structures. He received high recognition for his brilliant contributions: winning the Nobel Prize for chemistry in 1982; becoming the first South African to be president of the Royal Society of London from 1995 to 2000; being knighted by Queen Elizabeth II, and subsequently elevated to the Order of Merit.

I entered the school in January 1939 (in South Africa, the academic year follows the calendar year). In that first year, I made many new friends, some of whom have remained lifelong chums. I shared a desk with Peter Herbert Bradley Maytom and our close friendship has survived for 65 years. Peter, who was later a medical graduate of the University of Cape Town, went on to become president of the South African Medical Association.

While I was at Durban High School, the Second World War considerably altered the ways of life in the city. South Africa's east coast was on the circum-Africa route for convoys of British and Allied ships. They were usually travelling from England, but could not use the Mediterranean route to the Middle East theatre of war as they would have been extremely vulnerable to German dive-bombers. Consequently, they were forced to circumnavigate Africa. The great seaport of Durban was a convenient stopping-point for the convoys to take on supplies. Some of the convoys were so large that many ships could not be accommodated within the secure, landlocked Durban Bay with its berths and repair facilities. In these years I would occasionally count up to 30 or 40 ships riding at anchor in the roadstead. Great signposts everywhere conveyed the message, 'Don't speak about ships and shipping', and so I would simply spend time watching the vessels and imagining life aboard.

German U-boats and, after Japan entered the war, Japanese submarines preyed on the convoys. Some nights we heard gunfire and exploding depth-charges out at sea. I vividly recall the day in June 1940 when the war was brought home to our Durban skies. Two Italian cargo ships, the *Gerusalemme* and the *Jiulio Cesare*, were docked in Durban harbour. Churchill's *The Second World War* tells that on the afternoon of 10 June, the Italian Minister of Foreign Affairs informed the British and French ambassadors that Italy would consider herself at war with the United Kingdom and France the next day. The two Italian ships slipped out of Durban harbour and made a dash northwards towards Lourenço Marques (now Maputo), the capital of Mozambique (or Portuguese East Africa). Portugal was, of course, neutral during the Second World War.

In the morning – it must have been 11 June 1940 – my Form IVA classmates and I saw three planes of the South African Air Force, with bombs in position, fly low overhead and head off up the Natal coast towards Zululand in hot pursuit of the two Italian vessels. Three hours later we saw the planes return, their bomb-bays empty. One of the two ships had escaped to neutral Lourenço Marques, while the other was driven aground on the Zululand coast and the crew fled ashore. For a brief period they were at large. The radio warned citizens to take care and to inform the police if any of the Italians were encountered. As I recall, they were all rounded up in a few days, somewhat the worse for wear and rather hungry and thirsty.

Durban was a strategic city and the main road to Pietermaritzburg and the interior was a critical artery. The entire coastal strip and the area adjacent to the Pietermaritzburg road accordingly had a compulsory black-out. This affected public and private buildings, street-lights, and motor-car headlights. I became a member of the Civilian Protection Service in the suburb of Westville where I lived. My nightly duty involved bicycling around the neighbourhood inspecting whether any chink of light was showing through imperfect black-out curtains. I recall a satisfying moment that came one evening when I had to deliver a mild reproof to a defaulter who was one of my teachers.

At school, my peers and I had to learn how to operate stirrup-pumps (hand-operated water-pumps with a footrest that were used to extinguish small fires), while the smaller boys went into the large underground air-raid shelter with the matron. When alerts and alarms were sounded, classes broke up and everybody had to go to his appointed place. Despite many drills, we never came under enemy fire, something that disappointed my schoolboy imagination.

I CAN TRACE some of the seeds of my most important future interests back to these Durban years as a teenager. As I've already mentioned, I lived out of town in the suburb of Westville with my mother and Bert Norden, my godfather who had become my stepfather on 25 February 1937. I often had several hours to while away after school, as the bus service was not frequent. On other days, I waited for Bert to give me a lift home. During these periods my regular port of call was the Durban Natural History Museum, housed in the City Hall complex. On my lonely meanderings in the Museum, there were three exhibits in particular that fired my interest.

The first was on genetics and it featured displays from an hereditary point of view on human eye colour, 'mealies' (maize cobs) of different colours, and so on. This interested me greatly, especially given my sister's and maternal grandmother's deaths from diabetes. I suspect that it was here that the idea was implanted that led to my devoting my subsequent Bachelor of Science, Honours and Ph.D. projects to the study of chromosomes, the material basis of heredity.

The second exhibit, the Campbell Collection, was a wonderful display of South African mammals and other vertebrates. The enormous variety of these creatures from southern Africa, each identified by species and common name, geographical and ecological situation, fascinated me. Although the term biodiversity was not yet in common use, the Durban exhibit was a three-dimensional catalogue of tropical and sub-tropical biodiversity. I was later to learn that it was among the ancestors of such animals that early phases of human evolution were played out in Africa.

The third exhibit that opened my mind was archaeological in nature. E.C. Chubb, the curator of the Durban Museum, with G.B. King and Albert O.D. Mogg (who was later known to us at Wits as Uncle Bertie and who published a little book on the present-day flora of the Sterkfontein area), had excavated a cave deposit near the mouth of what was then called the Umgazana River on the coast of Pondoland or southern KwaZulu-Natal. The Museum displayed a number of the artefacts that had been recovered from the three-metre deep deposit. They were shown in a way that even a youth with no prior knowledge of archaeology could understand. I was captivated by the sheer wonderment of the world the exhibit revealed. It was my first revelation that there had been a human population in South Africa well before the coming of the white settlers in the seventeenth century and even before the arrival of Zulu and other Nguni-speaking peoples who live in that area today.

When I was not possessed by the Museum, I was downstairs in the Municipal Library browsing hungrily and avidly devouring book after book. Although a slow reader, I always had an insatiable appetite for reading. Family lore has it that I was already reading newspapers at three and four years old (although I am not sure that I always understood what I was reading as my dictionary habit came a little later) and my reading tastes were catholic. I have often told my students never to let a strange or unknown word pass before their eyes without looking it up in whatever dictionary they prefer to consult.

I was especially impressed by Chapter 37 in the Book of the Prophet Ezekiel. This is what it says:

> The hand of the Lord was upon me, and carried me out in the spirit of the Lord, and set me down in the midst of the valley which was full of bones,

> And caused me to pass by them round about: and, behold, there were very many in the open valley; and, lo, they were very dry.
>
> And he said unto me, Son of man, can these bones live? And I answered, O Lord God, thou knowest.
>
> Again he said unto me, Prophesy upon these bones, and say unto them, O ye dry bones, hear the word of the Lord.
>
> Thus saith the Lord God unto these bones: Behold, I will cause breath to enter into you, and ye shall live:
>
> And I will lay sinews upon you, and will bring up flesh upon you, and cover you with skin, and put breath in you, and ye shall live; and ye shall know that I am the Lord.
>
> So I prophesied as I was commanded: and as I prophesied, there was a noise, and behold a shaking, and the bones came together, bone to his bone.
>
> And when I beheld, lo, the sinews and the flesh came up upon them, and the skin covered them above: but there was no breath in them.
>
> Then said he unto me, Prophesy unto the wind, prophesy, son of man, and say to the wind, Thus saith the Lord God; Come from the four winds, O breath, and breathe upon these slain, that they may live.
>
> So I prophesied as he commanded me, and the breath came into them, and they lived, and stood up upon their feet, an exceeding great army.

I find it interesting that long before I ever knew of my future as a professor of anatomy I lingered over this passage and returned to it again and again. I also had no idea that one of my major research preoccupations throughout my life was going to be wandering in a valley of bones and trying to make these bones live – in my mind's eye to clothe them with ligaments and tendons and muscles, to see how they functioned, how their joints worked, how they stood and walked and ran. If I had my way, every laboratory of physical anthropology would have a framed copy of Ezekiel Chapter 37 hanging on its wall.

Bert Norden suggested that I should start a personal florilegium or commonplace-book in which I wrote down every new word I read, together with its meaning. I added choice phrases and memorable quotations that I came across. Amazingly, I still have the ample volume I started when I was fourteen. Even now, when time and energy allow, I like to add new entries to it.

THESE DURBAN YEARS saw the beginnings of my Jewish education. Although my parents were not deeply religious in the sense of observant, there had been a long history of involvement of the London Vallentines with Jewish affairs. My great-great-grandfather, Isaac Vallentine, was born in Belgium in 1793. Shortly afterwards, his family moved to London where they established a bookshop, soon to metamorphose into an Anglo-Jewish publishing house. Apart from publishing Hebrew and English prayer-books, Isaac founded the earliest Anglo-Jewish magazine, which gave rise to the *Jewish Chronicle* of London. Despite being bombed twice during the German 'blitz' on London in the Second World War, the *Chronicle* has continued publishing since its inception in 1841 to the present day. This record makes it the oldest extant Jewish periodical in the vernacular. Isaac also played a prominent role in the founding of the Jews' Orphan Asylum and other charitable and literary institutions in London.

Against this family background on my father's side, I was always drawn to Judaism. My very first formal education came in 1932 when I was enrolled at six years of age in a newly founded Jewish kindergarten, the Durban United Talmud Torah ('pupil of Torah'). Sadly I did not remain for more than a few months. I was having symptoms of what was vaguely described as a 'nervous breakdown', which kept me away from the kindergarten for the rest of 1932.

I was deeply involved in a Durban Junior Hebrew Congregation whose services were held every Saturday morning after the end of the main service. I served as president of the Junior Congregation for eighteen months. During that time I made up my mind to study for the rabbinate after leaving school. I paid a visit to Rabbi A.H. Freedman, to discuss how to set about this goal. It was a deeply disappointing interview. I discovered that the rabbi, who had come to Durban from Ottawa in 1937, was going through a period of intense disenchantment with the ministry and his congregation. His negativity came through poignantly to me and he talked me out of my idea of becoming a rabbi. When I left his apartment, I shed bitter tears of anguish, the last such outburst of grief of my boyhood.

Religion and my Jewish affiliations have nevertheless remained a vital part of my consciousness throughout my life. For over half a century I have remained a member of the Great Synagogue, Johannesburg, serving for a brief period on its Internal Management Committee.

THERE WAS ONE other respect in which my high school years in Durban laid the foundations of my future life. My inability to say 'No' when pressured into one responsibility after another had early teenage beginnings, and has been a feature of my whole life. At school I was made a prefect. This kind of appointment was usually confined to the leading sportsmen, but the school decided to make an exception in my case, acknowledging my strengths in the academic and leadership spheres. At the same time, I was Orderly Sergeant of the School Cadet Corps (handing out carbines from the armoury and typing all the Commanding Officer's letters for him). In addition, I was at various points the editor or co-editor of the school magazine; class captain; dux of the school; and, as already mentioned, a runner in the Civilian Protection Service. These activities were all over and above my studies.

As I was leading so many different lives at once – what nowadays would probably be called multi-tasking – it was inevitable that something had to be sacrificed. My varied responsibilities, as well as the fact I lived out of town and wartime petrol rationing limited mobility, meant that the social activities of youth, such as class parties and Saturday night dances, largely passed me by. The joys of youth, as sung by the poets and novelists, were elided from the springtime of my life.

Despite this suburban isolation, I was certainly no recluse. I had numerous close friends in Durban: my memory develops images of Raymond L., Michael L., Leon L., Percy C., Ramon L., Bernard K., Gerald L., Peter M., and enough others to make up a cricket team. I had some special girlfriends whom I worshipped, usually from afar. The beauty and sweet nature of some whom I put on a pedestal, including Shirley M., Rosemary W., Sybil L., Bernice J., Esmé I. and Joy L., overwhelmed me.

Fortunately, distance did not rob me of other and compensating kinds of fruition and satisfaction. Gladness could be felt through the reading of it; the romantic in me could react with empathy to Wordsworth's soliloquy on the daffodils:

For oft, when on my couch I lie
In vacant or in pensive mood,
They flash upon that inward eye
Which is the bliss of solitude;
And then my heart with pleasure fills,
And dances with the daffodils.

I even developed a little private philosophy – that happiness (for me at least) never occurs in the present tense, except in fleeting moments, but is to be plucked, like choice grapes or litchis, from an old orchard of memory. That was my theory of happiness in retrospect, in the past tense – and an especially delicious and rich flavour of happiness it was too.

4 Into the arms of medicine

As the end of my schooldays approached, I toyed with journalism and creative writing as possible directions for my future. However, my family, very sensitive to the hard times we had lived through, dissuaded me from such pursuits. Instead, I applied for admission to the Wits Faculty of Medicine.

My close friend Peter Maytom tried hard to get me to join him at the Cape Town Medical School, but I wanted to be close to my father who lived in Johannesburg. Since my family had broken up, I had been at school in Durban with my mother and stepfather, while spending vacations with my father. Now it was time to reverse the arrangement: to study in Johannesburg and holiday in Durban.

With a reasonable matriculation pass behind me (distinctions in English, Afrikaans, Latin and History), I was readily accepted by the Wits Medical Faculty at the age of seventeen. I have subsequently spent all of my academic and professional life at Wits, with never a moment's regret.

Save for their encouragement and pride, my parents played no active part in shaping my future. Neither of them had been to university. On my father's side, I was the first member of the family to attend university. On my mother's side, only an older cousin, Vincent Rosendorff, had studied at university.

I have always been deeply grateful that my parents were tolerant of my life-choices. When I desired to branch off temporarily from medicine, to take the Medical Bachelor of Science (B.Sc.) course, and a year later, the B.Sc. Honours, they let me have my head, although privately they thought I was quite mad adding to an already long (six-year) course of study. However, they put no obstacles in my way, willingly and affectionately abiding my foibles. They always rejoiced over my successes, though not always understanding what they meant.

FROM THE BEGINNING of my time at Wits I was drawn to the study of living things and especially loved the zoology course. The Zoology Department was headed by the Netherlander Cornelius van der Horst. His lectures in animal embryology fascinated me. In the manner of the day, he filled square metres of chalkboard with multicoloured drawings and diagrams. The gradual unravelling of the embryo to generate the anatomical finished product was little short of marvellous to me. These developmental messages were reinforced when I entered the Anatomy Department the following year.

Christine Gilbert (a Van der Horst protégée) and Joseph ('Joe') Gillman who carried out researches with Van der Horst on the embryology of the elephant shrew, were my embryology teachers in the Anatomy Department. Their contribution to this part of my formal education was formidable and cemented my passion for the study of development.

Later, when I became a staff member of the Anatomy Department, I presented a series of tutorials in human embryology to medical, dental and therapy students. It was to become one of my favourite topics for teaching. Still later, I developed a course of applied embryology lectures, with clinical applications. To this I added a course on teratology (the study of congenital abnormalities). This advanced course was especially for the training of postgraduates preparing for surgery, psychiatry and radiology. The course in clinical embryology and teratology drew undergraduate students and postgraduates from far and wide.

Another first-year influence was Professor John Phillips, who was the head of the Department of Botany. To our medical student class he gave a special programme on ecology. It was based largely on his own work on the control of the tsetse fly, *Glossina morsitans*. This fly in tropical and subtropical Africa is the carrier of the sleeping sickness parasite that affects humans and, in a form known as nagana, cattle. It had penetrated as far south as Zululand and northern Botswana, and I encountered the tsetse fly in Zambia during my researches on the Tonga people in 1957 to 1958 (see Chapter 23). Phillips's researches in East Africa and Zululand showed that the key to the control of the spread of the tsetse fly was through its ecology and interactions with animals and people. His important contribution to this field of study brought ecology firmly into my consciousness (and doubtless that of my fellow students too).

Years after Phillips's course, I was to make the first ecological study of the Kalahari San or Bushmen in 1951 (see Chapter 10). I have never ceased to wonder at, and appreciate, how many of my later interests could be traced back to the impact of greatly talented mentors and professors. Sadly, it seems that all too few of my contemporaries and my students acknowledge how much they owe to such dawnings.

Avid as I was to learn something about heredity, following my sister's death from diabetes, there was little or no formal genetics teaching in the first-year biology courses. Fortunately, my zoology textbook had a captivating chapter on genetics that added appreciably to what the Durban Museum had imparted to me. The magical world of chromosomes, the bearers of the genes, first swam into my ken through its pages and I was obsessed by chromosomes for more than the next ten years (with my Ph.D. thesis in 1953 and my first book in 1956 devoted to them).

IT WAS AT the beginning of 1944, my second year of medical studies, that I first set foot in the Wits Department of Anatomy. It was presided over by Raymond Arthur Dart, a larger-than-life personage (who is discussed in more detail in Chapter 29). When my class arrived, Dart was 51 years old and he had been head of the department for some 21 years. His fame and reputation as unpredictable, inspirational and eccentric had preceded him. Many earlier medical students had simply dismissed him as 'mad', but that was a very superficial judgement.

Dart's impact was largely the result of his unforgettable, exceptional personality. He was able to fire the most unlikely people with unbounded enthusiasm. He also brimmed with sympathy and the desire to help disabled people. When I was in second year, he devoted one morning a week to sessions with physically and functionally disabled students and citizens.

Dart was known as the Terror of the Dissection Hall. Unexpectedly he would enter the hall and give vent to a stentorian cry, 'Stop!' The class's activities came to a shuddering standstill. Then he would proceed to harangue us on his favourite subject: skill. When he permitted us to resume dissection, he wandered around the hall and, lighting on some awkward wretch, would snatch the scalpel and forceps from the student's tremulous hands and proceed to show him or her how it should be done. Once he was so angry with some fumbler that he emptied half a can of preserving fluid over his head!

Nowhere were Dart's histrionics more freely at play than in the steeply pitched Vesalian Lecture Theatre in the Old Medical School where I spent some 40 years of my life. Lecturing on the evolution of the upright posture and bipedal (two-footed) gait, he would lie prone on the elongated, marble-topped, lectern-table. Then he would demonstrate to us how the crocodile and other reptiles walked. Their legs were flexed sideways and to move forward their bodies assumed S-shapes, alternately convex to the left and to the right. Then came the great moment when mammals learned to lift their bellies from the earth on straightened limbs. From this quadrupedal position, Dart showed us how the upright posture of humans arrived. At this instant the dramatic high point was reached. The steep pitch of the central aisle in the Vesalian led upwards nearer and nearer to the ceiling as you climbed up the steps. A pipe ran longways suspended from the ceiling above the central aisle. On a certain day each year, Dart mounted these stairs with measured tread, without interrupting his flow of lecturing. There was a palpable tension among the students, many of whom knew what was coming. At the back of the auditorium, he turned around and clasped the ceiling pipe overhead. Then, still lecturing, he started to move forwards, hand over hand, his body and legs dangling freely over the retreating steps. At a certain point in his narrative, he cried in falsetto voice, '. . . and so they lightly came down to earth'. And he would come thundering down on to one of the landings between pairs of stairs. Whatever anatomy we failed to learn from Dart, we certainly never forgot how the apes (supposedly) came down to earth.

On another occasion, during one of Dart's orations in the Vesalian on the evolution of the erect posture, he focused on the tall Dennis ('Teddy') Glauber who was sitting on a stool taking notes on a pad on his lap, crouching with rounded back and shoulders. 'Ha!' ejaculated the professor in a rich, well-projected voice, his gaze directed at the inhumanly stooping student. As though he had received a shot of strychnine, Dennis straightened his back and sat bolt upright, closer to Dart's ideal posture. 'No, no,' cried Dart, 'Go back as you were.' Poor reddening Dennis resumed his malposture, assisted by firm pressure applied to the top of his head by the professorial hand. Once the whole class had seen this embarrassing spectacle on how not to sit, Dart applied one hand beneath Dennis's chin and the other to the top of his head and resolutely raised the luckless young man to his feet in a fully erect stance. Then, still supporting him, Dart wafted Dennis across the front of the room, his feet scarcely touching the floor. After this demonstration, neither Dennis Glauber nor any other member of the class could ever forget the difference between good and bad posture.

Dart's long-standing preoccupation with human posture and poise probably had its roots in the spasticity shown by his son Galen, as the professor recounted in a 1946 article, 'The Postural Aspect of Malocclusion'. Dart was convinced that he could educate the little spastic leg-muscles of Galen towards normal functioning. I first realised in 1946 how obsessed he was with little Galen's spasticity when Professor and Mrs Dart and their two children, Diana and Galen, accompanied two or three of us who were then Honours students, to Makapansgat, some 300 kilometres north of Johannesburg. He wanted to see for himself where the all-student expeditions of the previous year had collected fossil baboons and monkeys (see Chapter 6). These were to play a big part in the second research career of Dart and in the opening up of what proved in 1947 to be South Africa's fourth australopithecine (ape-man) site. After some hours at the now famous Limeworks, we returned to our campsite in the Makapan Valley. It was autumn, and the highveld trees and bushes shone gold and red in the slanting sun's rays. Before the drinks were brought out and supper prepared, Galen's hour followed. Dart persuaded Galen to walk, holding the child's heel and Achilles tendon down to the ground. 'Now walk,' said the great man, himself kneeling in the dust and clasping and directing Galen's heels. In this way he was trying to help his son's calf muscles from bunching up. Galen, with a frown on his brow, took a couple of steps with his heels down; then up they came as the spasm returned to the calves. Patiently Dart would start again with Galen's feet. Clearly this was a daily drill. Dart was convinced that by graded exercises he could cure, or at least greatly improve, Galen's stance and gait. How effective his ministrations were remains doubtful.

That evening we sat around the campfire. By lamplight I was developing with a needle a small mammal's lower jaw and teeth from the breccia (literally Italian for 'broken things'; rock composed of angular fragments cemented by finer material) which partially embedded them. Little Galen, bright-eyed and alert, watched my activities for a while and then asked me, 'What are you doing, Mr Bone-Picker?', a nickname that has seemed appropriate on many occasions during my career.

OVER THE NEXT 40-plus years, one of my research interests was the evolution of the human posture. This was partly due to Dart's influence and partly it followed my team's recovery from Sterkfontein of vertebrae, pelvic bones, hip-bones, knee-joints and ankle-bones (from 1966 to 1990), as well as the remarkable foot-bones of the specimen that Ronald John Clarke and I dubbed 'Little Foot' in 1995.

Many previous researchers on the attainment of the upright posture, such as Sherwood ('Sherry') Washburn of Chicago and Berkeley in the United States of America and John Robinson of Pretoria and Madison, Wisconsin, had concentrated on the pelvis and the hollow of the back, which undoubtedly played a very big part in the coming of the upright stance and gait. Not so much work had focused on the poise of the head and its relationship to the nervous system. In my studies, I stressed that the head of modern mankind is delicately balanced on a nearly upright spinal column. To attain this, the head and neck, including the brain and other parts of the nervous system, had undergone major adjustments.

Palaeontologists – since their business is old, fossilised bones – concentrate perforce on changes in the skeleton. Certainly, these are impressive and must have been pivotal in the achieving of uprightness and bipedalism (standing, walking and running on two feet). We dare not, however, neglect the role of the nervous system, that is, the brain, spinal cord and nerves. The musculoskeletal adjustments that effected efficient uprightness and bipedalism improved the balance of the bony framework. This in turn led to a situation in which we humans can stand erect with little muscular activity. While a gorilla needs powerful muscle contraction to stand upright even for a short time, a modern man or woman requires only delicate muscular contractions to maintain uprightness even for long periods. It was this realisation that directed my attention to the subtlety of the information that the sensory nerves bring to the central nervous system, during such an apparently simple act as bipedal standing or walking in a relaxed fashion.

Two great sets of sensory inputs are essential to the upright posture: vestibular and proprioceptive. The vestibular messages bring information about balance. This is a crucial part of standing upright. We do not readily lose our balance unless we are rather elderly, or diseased, or have imbibed too heavily, or are trying to keep our balance on a sloping or slippery surface or in the teeth of Cape Town's notorious southeaster. The vestibular messages have another important function: they tell us how our body and individual parts of it are orientated in space. Yet upright standing and walking are not simply exercises in balance.

The other vital sensory input comprises proprioceptive messages. Such messages reach the spinal cord and the brain from the 'anti-gravity' muscles, ligaments and joints. They set the precise muscle tone, the flickering contractions necessary to maintain erect posture and relaxed bipedal walking. Such sensory information doesn't apply only to the trunk and the limbs, but also to the head and neck in the upright posture.

The questions to which I dearly wish we had answers are these. If proprioception plays an important part in the control of the upright posture and the poise of the head, does it differ between humans and apes? If there are differences, what is their nature? Are there variations in the kinds of specialised nerve-endings, in their numbers and density, their distribution and patterns? What do we know of the pathways from these proprioceptive nerve-endings to the central nervous system? Are they differently, perhaps uniquely, developed in humans as compared with quadrupedal animals? Are these nerve-tracts denser from some areas of the body than from others? Are they more intricate, with more cross-connections, in humans than in apes?

We need much more evidence before these problems can be resolved. At this stage we have no ready answers to the question of whether and to what degree our success as bipeds, with gaze directed to the horizon, is to be laid at the door of the anatomical adjustments of our skeleton, and how much to a more exquisitely developed proprioceptive system. I do not expect to be on this planet when these questions are answered, but I have every confidence that they will be solved before the twenty-first century is much older.

5 | Getting to grips with genetics

It was the night before my oral examination in histology. I was to present the results of my project on mammalian chromosomes, and I was worried.

For my B.Sc. project, I had focused on the chromosomes of the albino Norway rat, a popular laboratory animal. The techniques I used in those days were primitive. Furthermore, it was most difficult to work on the chromosomes of mammals as they were very small and usually numerous. In my squash preparations I struggled to find good spreads of dividing cells. It is during cell division that the chromosomes become readily apparent. Using a crystal violet technique to re-stain my squash preparations, I suddenly found beautiful bundles of brilliantly stained structures. For a week or two I examined these and rejoiced at the sheer beauty of the entities that my technique had displayed.

However, nobody was available to confirm whether I was on the right track or not. My mentor, Joe Gillman, gave me the impression that he did not believe in the existence of chromosomes. Sydney Brenner was working on the chromosomes of the elephant shrew, but he was unavailable to me. My viva voce examination was drawing closer. I was concerned that my glorious serried ranks of chromatin bodies were too like one another, too homogeneous. If I had doubts myself, just imagine what the reaction of my examiners would be.

Working alone in the laboratory late on the night before the oral exam, I suddenly chanced upon plates of well-stained indubitable chromosomes. They showed the hallmarks of these DNA threads as seen in the middle stage (metaphase) of cell division. I let forth a great cry of jubilation in the otherwise deserted medical school. It was truly an eureka moment, one of the first in my career. (It subsequently emerged that what I had previously been assiduously counting were the heads of sperms – no wonder they had looked homogeneous!)

PERHAPS I SHOULD explain that after my second year as a medical student I had decided to become a Medical B.Sc. student, majoring in the Anatomy and Physiology Departments. The Faculty of Science accepted the first two years of the medical course as tantamount to the first two years towards a Bachelor of Science degree. One further year of advanced courses would give successful students a B.Sc. degree.

I was required to take one minor and two major courses. The Physiology Department offered one major course, while the Anatomy Department gave a choice of two: Anatomy with Anthropology, and Histology and Embryology. The physiology major was compulsory and so I had to choose one or the other major in the Anatomy Department. Already I was gripped by anthropology and at the same time my love affair with genetics and chromosomes was powerful. I felt it was too soon for me to be compelled to make a choice between the two anatomical majors. So I requested permission to do a conjoint major in anatomy. My petition was denied by Dart and Gillman during a lengthy and rather painful interview, and I was obliged to choose. I decided on the Histology and Embryology option, but let it now be whispered, I spent a good deal of time in that year unofficially doing anthropology which was a big part of the other anatomy major.

In addition to the two majors, we were required to take a minor subject, History and Philosophy of Science. The course was offered by Brian Farrell of the Philosophy Department. He was a most dynamic teacher and had a strong effect on all of us. For the purposes of his course he adopted a logical positivist approach and adhered to it rigorously and consistently. Farrell touched on fundamental questions such as the existence of God and showed us how this question would be approached by logical positivists. Some members of our class suffered sea changes as a result. Those who had come from a religious background found their cherished beliefs challenged. One student who was the son of a Dutch Reformed Church elder abandoned his faith; a son of Baptist missionaries changed to the Methodist Church; someone suffered an anxiety neurosis. Everyone was forced to think. In my science class at medical school, we responded by setting up a Comparative Religion discussion group, for among the thirteen of us, eleven or twelve faiths, denominations or personal philosophies were represented. Worried about the seeming incompatibility between religion and science, especially evolution, a group of us went to visit the Revd Joe Webb of the Central Methodist Church in Johannesburg. We asked him along what lines we might set out to find reconciliation between religion and science. He did not believe that these were irreconcilables and encouraged us to keep searching. This I have tried to do all my life.

MOST STUDENTS WHO had branched off from medicine – there were thirteen of us in 1945 – returned to their place in the medical class after the year's interruption. Four others, myself included, registered for a B.Sc Honours degree. Although my father could not afford to pay for the extra year, I was fortunate, as in the past, to be able to win bursaries and scholarships that allowed me to pursue my studies.

In my B.Sc. Honours year, I pushed forward my work on chromosomes. At this stage, with improved techniques, I based my studies on the gerbil. This little rodent, whose scientific name was *Tatera brantsii draco*, is a carrier of plague in southern Africa. I caught my specimens at Bronkhorstspruit east of Pretoria.

It was a full and exciting year, replete with laboratory 'discoveries', wide and deep reading, writing up of the previous year's expedition to Makapansgat Limeworks (see Chapter 6), as well as further field excursions to Sterkfontein, Kromdraai, Makapansgat, Gladysvale and Bolt's Farm. I participated for the first time in a systematic archaeological excavation at Rose Cottage Cave close to Ladybrand in the eastern Free State, under the direction and guidance of 'Peter' van Riet Lowe, the Secretary of the Historical Monuments Commission, and his second-in-command, 'Berry' Malan, a meticulous excavator. All the time, having been frustrated in my attempt to major in both anatomy options, I was trying to synthesise the two thrusts, genetics and evolution. It was clear that the genes and their mutations were the building stuff of evolution, but how did they work?

During this year, I said farewell to my teens and embarked all agog on my twenties. I marked this rite of passage by reading two major works from cover to cover. The first was Charles Darwin's *The Origin of Species* (1859). This work is, of course, a landmark in the study of evolution. It provided such a wealth of evidence that change had occurred over past aeons that the acceptance of evolution became irresistible. In addition, Darwin gave us a mechanism, natural selection, by which evolutionary alterations could have occurred. The human mind takes more readily to a new idea if it understands how the new notion could work.

The other book in which I became absorbed at that time was *Genetics and the Origin of Species* (1937) by the Russian-American Theodosius Dobzhansky ('Doby'). His writing made me realise that it was not a forlorn hope to believe that genetics and evolution could be synthesised. He showed how genetical variation was the material basis of evolution (to borrow Richard Goldschmidt's phrase). It must be remembered that Darwin had known nothing of genetics and had even developed some of his own rather quaint ideas on heredity.

Genetics and the Origin of Species crystallised my bigamous affair with cytogenetics (the study of the chromosomes) and palaeo-anthropology (the study of humankind's past). Doby became one of my heroes (see Chapter 30), and I avidly digested as many of his writings as I had time to do.

Joseph Needham was another writer whose work I was introduced to at this time. Needham was an extraordinary polymath – a man who was an outstanding scientist, especially in the fields of biochemistry and morphogenesis, a Christian (High Church) by conviction, a Marxist, an authority on the science and civilisation of China, a philosopher and an historian.

His three-volume book *Chemical Embryology* (1931) and his great encyclopaedic monograph on *Biochemistry and Morphogenesis* (1942) became essential reading for our courses in experimental embryology, especially the sections on concepts. In addition, Needham's personal synthesis helped some of us to realise that science and religion, although they seemed to be diametrically opposed to each other, were reconcilable.

ONCE MY HONOURS degree was completed, I returned to the medical course where my place had been kept for me. At the same time, by special permission of the deans of the Medical and Science Faculties, I was allowed to register concurrently for my Master of Science degree (on a part-time basis). These were strenuous years for me: as well as being a Master's and a medical student, I was earning my keep (in part) by teaching in the Department of Anatomy; serving as Class Representative of my medical class; taking part in student affairs as an officer of the Students' Medical

Council and of the Students' Representative Council; and playing a vigorous role in keeping alive and building up Nusas, the non-racial National Union of South African Students (see Chapter 9). The pattern of simultaneous multifarious activities that had become apparent in my teens in high school, was now in full flood.

Throughout these years, from 1946 to 1949, my genetical researches were ongoing and, as my M.Sc. project flourished, the department and the faculty agreed to convert my status to that of a Ph.D. student.

When I was doing my cytogenetical research in the middle to late 1940s, there were pitifully few investigators working on the chromosomes of humans and other mammals. By correspondence I was in touch with all of them known to me. In Texas there was Theophilus Painter; in London, P.C. Koller (who invited me to come and work with him); whilst at Lausanne in Switzerland, Robert Matthey was a leading cytogeneticist (who offered me a job when I visited him in the 1950s). In Japan S. Makino was following in the footsteps of K. Oguma. The only other centre for such studies at that time was the Anatomy Department at Wits, where Sydney Brenner had been working on the chromosomes of the elephant shrew and I on those of the albino rat, the gerbil, the human species, Chacma baboon, chimpanzee and *Galago* (night-ape or 'nagapie' as it is known in Afrikaans).

Late nights in the laboratory were sometimes soporific and sometimes exciting. Often mine was the only light burning in the medical school. On such occasions only the nocturnal grunts and sometimes a cacophony of sex-cries of the Chacma baboons in the Anatomy Department's baboon colony on the top floor kept me company. Whenever I felt sorry for myself in my isolation I would comfort myself with the thought that at least they were fellow primates.

It was not, however, only the hairy kind of primate to which I was drawn. I was strongly attracted to one of my fellow Honours students, Pris K-S. From a little distance I fell deeply in love with her, but was vexed by the question of how I should or could reduce that distance. I was beginning to take steps in that direction in my own sensitive way, when I became aware that she and Mike W. were already united in love – and were subsequently to marry. Bah, I had been pipped at the post!

This feeling of rejection did not last long and I went so far as to become unofficially engaged to Ruth K. Her parents lived in the same boarding-house as my father and I did, so I saw her frequently. On Sunday afternoons we went off to The Wilds in Johannesburg and arm-in-arm we warmed to the atmosphere of the setting sun and the cool evening breeze: those were magical days. I took Ruth down to Durban to meet my mother and step-father. It was not a successful visit and my family moved heaven and earth to break off the relationship. Realising that Ruth wanted to get married soon, that I had more years of medical studies and a Ph.D. ahead of me – and above all that I had no money to speak of and certainly could not at that stage countenance assuming responsibility for her and myself – our informal engagement was broken off. The whole relationship had so shaken me that, for the first time, my exam results that year were below my usual standard – and it was the psychiatry exam, in particular, that produced a pass that was below par!

I returned to my nocturnal studies in the Anatomy Department, where my M.Sc. had metamorphosed into a Ph.D. in the making. My discovery of a new kind of cell division, that I called 'premeiotic mitosis', was one of the unexpected and fascinating results of my doctoral studies. I described it at some length in my 1956 book and it was also the subject of my first presentation to the Anatomical Society of Great Britain and Ireland in London in 1955, when I was spending a year as a Nuffield Senior Post-Doctoral Fellow at Cambridge University.

THE VIEW THAT scientific research should be pursued wherever it might lead went up in smoke, once and for all, on 6 August 1945, when the atomic bomb was dropped on Hiroshima. Of that moment J. Robert Oppenheimer, who is spoken of as the father of the atomic bomb, said, late in his life, 'In some crude sense, which no vulgarity, no humour, no overstatement can quite extinguish, the physicists have known sin.' The atomic bomb raised in an acute form the question of the interrelationships of science, society and ethics.

None of us who were young germinating scientists at that time could escape the news from Japan and the aftermath. I believe that 'The Bomb', as we called it then, marked a turning-point in the history of science.

Throughout my life's work, I have been conscious of the problem of the nature, role and social function of science. With these questions is coupled the riddle of ethics in relation to the scientific endeavour. There is not only a philosophical but also an historical aspect to these enquiries.

Until the Second World War, the scientific search for the truth was considered to be a mission for the good of all, to be shared whether there was a war on or not. As Jacob Bronowski reminded us, the roots of this attitude went back to the 1660s when the Royal Society (of London) and the Académie des Sciences (of Paris) were founded. From then until the end of the Napoleonic Wars, France and Britain were at war for a total of 60 years. Yet, during all that time, scientists in the two countries freely exchanged information, visits, letters and awards. For example, when Edward Jenner discovered the vaccination against smallpox in 1798, the news circulated freely on both sides of the channel, although the two countries were engaged in a 'hot war'.

Between the two world wars, most scientists – if they thought about the matter at all – believed fervently in the pursuit of the truth, wherever it might lead. It had a certain holiness about it, this quest for the truth. It even provided many scientists with a credo, a set of values and beliefs. Science shared with religion this respect for the truth.

In this faith I and my generation, as young students, grew up believing that 'The truth should be pursued whithersoever it may lead'.

Scientists, it seems, were willing to leave their discoveries and inventions like foundlings on the doorstep of society (as Richie Calder happily phrased it). What happened to their discoveries, it was held, was for the politicians to decide.

Even then, there were two schools of thought. Some people were of the opinion that the spirit of science was exclusively dispassionate, objective, cold and aloof. They believed that science recognised no *ethikos* and no *aisthetikos*. Questions of ethics and of aesthetics, these scholars believed, resided in the realm of philosophy, not of science. For others, such a definition was a purist's distillation of the essence of science, a caricature that was too other-wordly, too absolutist, too essentialist. They pointed out that science without an ethic might produce such aberrations as racism (and a supposed scientific underpinning for its political manifestations such as

Nazism, apartheid and the latest horror, 'ethnic cleansing'). Science without ethos could give us negative eugenics, biological warfare and nuclear weapons. For this group of scientists, it was irresponsible to exempt or exonerate science from the ethical consequences of its discoveries.

Hiroshima and Nagasaki changed those earlier, almost naïve notions and scientists were now forced to take note of what happened to their discoveries. This was true not only for the physical sciences, but also for the biological sciences. Reflect for a moment, if you will, on such bio-ethical questions as the human genome project, stem cell research and genetic engineering, including the cloning of individuals. No longer could the scientist shrug off the challenge by declaring pitifully: 'I am just a scientist: it is the politicians who put my discoveries to good or bad use. What can I do?'

In two dramatic moments, the role of scientists had been remoulded and would never again be the same. It was realised that science could not and should not be neutral and that scientists were morally obliged to take heed of where their work might lead. Thus, in my lifetime, the whole faith of scientists has changed. Fifty years later I sat alone in the Peace Park at Hiroshima, ruminating on these matters. It had been verily a revolution of ideas that science, and I as one of the latest neophytes, had lived through in 1945 and the years that followed.

6 The lure of Makapansgat

One of the most scenically lovely of the South African fossil ape-man sites is Makapansgat. It lies just to the east of the Great North Road between Potgietersrus (now Mokopane) and Pietersburg (now Polokwane), some 300 kilometres north of Johannesburg and roughly halfway between there and the Limpopo River.

I first visited the area as a nineteen-year-old in 1945, accompanied by another member of my Medical B.Sc. class, Joseph Stokes Jensen. On that reconnoitre, we made the appropriate contacts and arrangements for a winter expedition that was to follow shortly.

There had been a long tradition of third-year science-class initiatives. In 1924 a member of the first such class, Josephine Salmons, had borrowed from her fellow student 'Pat' Izod, a fossil baboon that his father Edwin Gilbert Izod had brought to Johannesburg from Taung. This unleashed a chain of circumstances leading to the recovery of the renowned Taung child skull in November 1924.

In 1936, another of Dart's students, Trevor Jones, recovered cercopithecid (baboon and monkey) skulls and endocranial casts from George Barlow who had obtained them from Sterkfontein. In the same year, B.Sc. student William Harding le Riche and lecturer Gerrit Schepers obtained more cercopithecids from Sterkfontein. When they took these to Robert Broom, another sequence of events was set afoot, leading to the recovery of the first adult *Australopithecus* nine days later, on 17 August 1936 (see Chapter 8).

A similar train of events started when, in July 1945, members of my own science class found the first baboon and monkey fossil specimens at the Makapansgat Limeworks. This led directly to the recovery of the first hominid specimen by James W. Kitching 26 months later.

IT WAS ALMOST by chance that we chose Makapansgat for our winter expedition. We first consulted John Phillips, head of the Wits Botany Department, and he suggested the Zoutpansberg as the most suitable spot. When we found from our atlas that this was nearly 650 kilometres from Johannesburg, we presumed that the rail fare would be exorbitant. Halfway, approximately, lies Potgietersrus, which seemed to be an alternative venue. Many people were encouraging about our proposed expedition, giving us advice and lending the equipment that we would need. Professor Dart, in particular, was favourably disposed: he reminded us that it was exactly twenty years since he had first drawn attention to the significance of the Makapansgat Limeworks as 'a site of early human occupation'.

Our expedition lasted from 28 June to 8 July 1945 and was a multidisciplinary excursion. Apart from the work of the geology, botany and zoology students, some of the medical science students spent their time breaking up collapsed breccia at the Cave of Hearths and the Rainbow Cave, and collecting the archaeological objects laid bare. From the Rainbow Cave breccia, I recall recovering a superb circular scraper.

Every evening around the campfire, one or other person gave an informal talk about some aspects of our personal field of expertise. Tony Allison spoke about the birds of the valley (when he had made his survey of them); Johnny Kirkness told us about the rock formations of that area; Joe Jensen and I threw out a few tidbits on the archaeology of the Cave of Hearths and Rainbow Cave and the story of the Historic Cave. The pièce de résistance was Brian Maguire's modest, quietly-spoken chat about the vegetation of the area and the varied ecological zones from the floor of the valley up to the summit over the dolomites and, at the head of the valley, over the grey quartzite cliffs, the nickpoint of the stream that runs through the lush valley.

On 2 July 1945, the Medical B.Sc. students who were working at the Makapansgat Limeworks recovered the first primate from the Limeworks: there were a cranium, endocranial cast and, close by, a mandible. The jaw was subsequently made the type of a new species, *Papio darti*, named in honour of Raymond Dart. Later the species was 'lumped' into a different genus as *Theropithecus darti*. Other fauna that we recovered from the Limeworks included equid and bovid specimens.

Robert Broom, a palaeo-anthropologist who lectured us on a part-time basis in the Anatomy Department, was very pleased with what we had achieved and was especially impressed that we had found a new type specimen of a cercopithecid, that is, a member of the family *Cercopithecidae*. As class representative and expedition leader, I wrote to Broom on 8 August 1945:

> We were all very encouraged by the favourable manner in which you viewed our recent expedition, and the several finds which were made. You have more than kindled a spark for the future, in several of us, and I am sure that your work on anthropology and palaeontology will be taken up and pursued in this fruitful field of South Africa.

In reply Broom wrote to me on 10 August 1945:

> . . . I saw enough [of your Makapan finds] to know that you are on a good wicket. A brief look over your fossils showed that you have two skulls of Primates which when cleaned out may prove to be new and may help us to date the other fossils. You have one or two antelopes which should be identifiable at least generically. You appear to have the tooth of [a] new animal quite unknown to me. All this shows you have been working in a bed of an age different from the Kromdraai and Sterkfontein deposits. In fact you are exploring a Terra Nova. Quite likely in the same deposit you may find some new sort of early man or pre-man which may be even more important than the discoveries made at Taungs [*sic*] or Sterkfontein or Kromdraai.

On 27 September, Broom followed this up with a letter to our lecturer in the Anatomy Department, Alexander ('Sandy') Galloway:

> You might tell the students that they have made a discovery of considerable importance. The 'baboon' they found at P.P. Rust [Potgietersrus] appears to represent a new genus. Probably it is also a new species. I have named a fossil 'baboon' *Parapapio major* from Kromdraai, but this is only known from a few upper molars. Of the P.P. Rust specimen we only have lower jaw teeth. Still I feel almost sure that it will turn out to be new. Certainly it is not a *Parapapio*.

> Please ask them to do no further development till I see them – as they may do harm. Of course the beast is clearly a Cercopithecid . . .

Our student expedition to Makapansgat in June and July 1945 had important consequences. Our revelation of extinct cercopithecids at the site kindled in us the high expectation that hominids would follow. We were well aware that, at both Taung and Sterkfontein, the recovery of cercopithecids had presaged the discovery of hominid specimens. (And, indeed, as already mentioned, an australopithecine specimen was discovered at the Limeworks by James Kitching in September 1947.)

The presence of extinct species of baboons and monkeys had become an almost magical marker of the presence and discovery of fossil hominids. We were not at all sure why the two kinds of primates should be so closely associated. There were a number of speculations, most striking being that of Dart who believed the hominids killed and ate the baboons. Later Charles Kimberlin ('Bob') Brain produced very convincing evidence that both the hominids and the cercopithecids were killed and eaten by carnivores, especially the big cats, and it was Louis Leakey who suggested that the big cats liked the taste and smell of primate flesh, both hominids and cercopithecids.

Another consequence of our first expedition to Makapansgat was the flaring up of a feud that had simmered between the Anatomy Department and Robert Broom in and after 1936 (see Chapter 8). The fact that we were students from Dart's Anatomy Department, but that Broom (with Jensen) published the first identified Makapansgat fossil baboon in 1946, certainly generated some tension, which echoed the hard feelings of 1936.

Six more expeditions were mounted by the Anatomy Department in the ensuing eighteen months. In September 1945, I led a team on our second expedition to break up breccia and collect fossils at the Makapansgat Limeworks. On our third expedition in December 1945, we recovered a number of additional baboons and monkeys. Our finds in December 1945 included the type specimen of a new genus and species. It was described by one of the anatomy Honours students, Ollerais Mollett (1947), and named *Cercopithecoides williamsi*, after Eric Williams who recovered it. For several years it remained something of an enigma in the Anatomy Department. Then, on a visit to Johannesburg, Louis Leakey examined it and declared that it was a colobine. These tree-dwelling monkeys are known today in East Africa, but not in South Africa. Yet, if Leakey was correct, it would appear that the colobines had lived in South Africa millions of years ago, and then died out in the southern African subcontinent. In East Africa, however, they survive to the present day.

Leakey's claim that *Cercopithecoides* was essentially the same as the living African colobines was at first resisted, for example by Leonard Freedman, our leading authority on southern African fossil baboons. Later, on careful comparative studies of more material, it has come to be accepted that Leakey was right and that *Cercopithecoides* was one of the genera of the Colobinae (that is, the sub-family to which the *Colobus* monkeys were ascribed).

ALL OF THESE names and debates might seem confusing. Some of the fiercest arguments in anthropology have revolved around what names to give different fossils. Taxonomy is the field of science that deals with the rules for the classifying of natural objects and for the naming of species and other groupings of living things.

There are different philosophies or approaches to the relationships between varied groups of creatures and to the naming of these groups. Some scientists seek out differences. They are 'splitters', and they tend to multiply the numbers of species, sometimes excessively – to the great confusion of students and even researchers in the field. By contrast, some seek out similarities and base their classifications on these; they are 'lumpers' and they tend to minimise the numbers of species. Some people chop and change.

When Dart erected a new species, *Australopithecus prometheus*, for the Makapansgat hominid, he was splitting; when, however, he proposed to place all of the ape-men (including those that had been named *Australopithecus, Plesianthropus* and *Paranthropus*) into the single genus, *Australopithecus*, he was lumping. Of Robert Broom, I think it may fairly be said that, throughout his life, he was a splitter and he created scores of new species and genera.

There is no absolute right or wrong about lumping and splitting. Which you choose may depend on your personal philosophy of evolutionary change, or it may depend on your temperament. Personally, I have always inclined to the lumping philosophy. For one, there is a scientific basis for the lumping of groups of extinct or living things into fewer but larger units. A greater awareness of variability is likely to be connected with a tendency towards lumping (and hence fewer species and, at a higher level, fewer genera).

Lumping leads to a simplified picture of nature, but it is founded on more than just a penchant for simplicity. There is a philosophy behind it, the concept of wholeness, of holism. Perhaps it reflects also my quest for synthesis, which has been with me all my life. This concept I tried to apply to my lifetime of research: seeking synthesis between genetics and evolution, between the anthropology of the living and of the dead, between science, history and philosophy, and fundamentally, between the scientific and the spiritual realms.

IT COULD BE argued that the most important consequence of our student expeditions (six in total) was their effect on our professor, Raymond Dart. For many years, since the exciting – and depressing – events of the 1920s, when Taung was revealed by Dart and widely rejected by most of the scientific world (see Chapter 29), Dart had played very little active part in palaeo-anthropology. When asked if his silence in this field was a response to the extraordinarily negative treatment he and his claims had received at the hands of most European and American scientists, he would deny this angrily. He was too busy; he had many other things to do; he was leaving it to Broom and the younger scientists in the Anatomy Department.

The role of the first student expedition in coaxing Dart back into palaeo-anthropology is best recounted in Dart's own words:

> Then in 1945 a student adventure led by a member of my science class, P.V. Tobias ... was responsible for thrusting me back into the maelstrom of man's beginnings. Tobias and his group, thirty strong [actually 21], visited the Makapansgat Valley ... and brought back a story so astonishing that I could no longer resist returning to that earlier field of activity ...
>
> When the party returned they said that the two sites [those from which stone tools and fossil ash came and the source of Eitzman's 1925 bones] were a mile apart. There were many Old Stone Age implements at the Cave of Hearths and plenty of Middle Stone Age implements at the Rainbow Cave. However, at the limeworks site a mile lower down the valley – where the fossil-bearing gray breccia was found – there were no

traces of implements. But they had brought back something much more significant than Stone Age tools. Among the grey breccia they collected was the skull of a fossil baboon, indistinguishable from the *Parapapio broomi* first found by Trevor Jones at Sterkfontein in 1936 and now recognised as a characteristic of australopithecine deposits.

'Doesn't this mean, sir,' Tobias asked, 'that Makapansgat may be far older than you or anyone else imagined?' 'It does indeed. It certainly looks that way,' I replied. 'Then,' asked the young student, 'doesn't this tempt you back into the field of anthropological research?' Hesitantly, he added, 'It might even prove to be contemporaneous with Sterkfontein.'

It was almost as if he had read my thoughts. It might not only prove to be as old as anything yet discovered but might also yield a more complete man-ape than those found by Broom. Summoning Tobias to follow, I went to my workshop and took down my hammers, chisels and other anthropological tools which had lain neglected for so many years. 'You have my answer,' I told him.[1]

Dart's return to palaeo-anthropology was marked by the revelation of a series of hominid specimens from the Limeworks, recovered for Dart by James Kitching of the Bernard Price Institute and his brothers Ben and Scheepers, and by Alun Hughes of the Anatomy Department. It produced also Dart's ingenious but flawed concept of the Osteodontokeratic Culture (bone, tooth and horn culture), which he supposed had been practised by *Australopithecus*. The work at Makapansgat gave Dart a major new scientific career between 1947 and 1966, beginning some twenty years after his earlier meteoric phase with the Taung child.

All in all, some 30 or 40 hominid specimens have been recovered from the Makapansgat Limeworks. Although small in quantity, as compared with what has come from Sterkfontein, Swartkrans and Drimolen, in quality the Limeworks hominid fossils, a good mixture of juvenile and adult specimens, are of superlative texture and finish, and have enabled us to explore and answer some important questions.

HOW DID ALL of these exciting happenings and results of the mid-1940s affect me? I was the *primum mobile* in the organising of the expeditions and related initiatives. Was there perhaps another consequence of these early student endeavours, which started when I was not yet out of my teens? An onlooker would doubtless claim that these undrilled and rather rough-hewn adventures played a seminal role in determining the direction of my own future career. Whether the events were cause or effect, it is hard to determine. I cannot help thinking that the die was already cast.

7 | Unravelling the past in Mwulu's Cave

Serendipity, coincidence, synchronism, eureka moments – these have garnished my life-track beyond all expectation. The most striking serendipitous occurrence of my student years marked my first visit to the cave I was subsequently to name Mwulu's Cave.

Brian Maguire, who had accompanied us on our first Makapansgat expedition in 1945, was a remarkable, largely self-taught field botanist. He had been a science student at Wits University until he developed some behavioural oddities. One expression of these was his addiction to tea, of which he drank about '40 to 50' cups a day. It had taken a lot of persuading before Maguire agreed to come on the Makapansgat expedition, and I believe it was a turning-point in his rehabilitation (and he eventually came back to Wits and finished his degree).

In the course of a conversation, Maguire mentioned that, high above his father's old farms, *Spanje* and *Portugal*, a yellowwood tree (*Podocarpus latifolius*) had taken root in the cramped confines of a small cave. Unable to grow erect, it had been impelled towards the cave mouth, growing horizontally. It was an unusual botanical phenomenon and Brian said he would like to show the tree to me.

Maguire's two farms were at the foot of the massif that marks the eastern face of the highveld, about eight kilometres east of the Makapansgat Valley. When we arrived at the farms, one still inhabited by tenants, the other in ruins, we were faced with a fairly steep climb up the west face of the massif. After some serious leg-work we found ourselves in a wide neck of the mountains, opening up to a marvellous view of the lowveld more than a thousand metres below us. Straight ahead, to the east, was the countryside made famous by Sir James Percy Fitzpatrick in his 1907 book, *Jock of the Bushveld*. To the north, beyond the glinting roofs of Pietersburg (renamed Polokwane) nearby and Louis Trichardt (now Makhado) afar, towered the distant Zoutpansberg. From this elevated position, we commanded a view westwards to the peaks of the Waterberg. To the south lay the open plain of the Springbok Flats and beyond that, far to the south, though not visible by day, was the lofty pile of Pretoria (now Tshwane). Black eagles soared, not above us, but far below, while on some easterly buttresses we saw Chacma baboons at play. It was exhilarating and, coming on it so suddenly, I quite lost my breath.

Pausing there, drinking in the picturesque scene, it was all too easy to allow myself to become mesmerised by its sheer spectacle, but we had come for another purpose. 'The cave is up here,' Brian urged gently, and we clambered up the right or southern shoulder of the neck. The final approach to the cave mouth was a vertical ascent and all of a sudden we had arrived.

There was the podocarp tree, growing in its cramped surroundings, rooted in the sandy floor of the cave filling. Brian pointed out the features from which he had identified the species. With the cave roof low over our heads, I was down on all fours to have a closer look at the tree's diagnostic traits as Brian expounded on them. In that position my hands were in the soft, sandy deposit of the floor. Suddenly I felt a hard object beneath the surface. I pulled it out and it was easily recognisable as a stone tool of an era long past. With its prepared platform and absence of secondary trimming on the flake surface, it would have been quite at home in the Pietersburg Culture (as we called it then), a phase in the southern African Middle Stone Age which went back some 50 000 to 100 000 years from the present. If there was one stone artefact just where my hand had alighted on the sandy deposit, it was very likely that there were more in the deposit into which the yellowwood tree had sent down its roots.

I subsequently organised and led a party, comprising Ollerais Mollett, Percy Barkham, Anthony Allison, Owen Jones and myself, to excavate the cave. Jones was a member of the Anatomy Department's technical staff and the other four of us had obtained our Medical B.Sc. and Honours degrees in the department. As already mentioned in Chapter 5, six months earlier I had learned the art and skill of making an archaeological 'dig' at Rose Cottage Cave in the eastern Free State, under the tutelage of 'Berry' Malan and 'Peter' van Riet Lowe. As a result, I was given a permit to excavate the Mwulu's Cave deposit. We camped on the leeward side of the mountain and each day clambered up to the cave with equipment, water and *padkos* (food for the road). Each evening we came down the precipitous descent with our carefully labelled bags of artefacts. From the cave mouth we could watch great storms growling ominously across the low-lying ground to the east, sweeping up the sheer face of the Strydpoort mountains – and then crashing over the top where the *klipspringers* (small mountain antelopes, *Oreotragus oreotragus*) held sway, or howling through the neck. If we had made it in time, we listened to the rain pounding on our canvas tent, each raindrop a teaspoonful of water.

Over 3 000 stone tools were recovered from three strata of increasing technical complexity from below upwards. It was the first archaeological excavation of a cave deposit that had been made in the former Transvaal province (this part of which is now the Limpopo Province).

You go to search for something – an odd tree – and you find something else, something that may prove to be even more important than that which you had set out to examine! This is the essence of serendipity.

THE CAVE HAD no specific name up to then, but Constantine Duncan ('Con') Maguire told me that an old Zulu recluse named Mwulu had been living there early in the twentieth century. Indeed, a few of his broken potsherds were still lying around on the surface of the deposit. For this reason I gave the site the name Mwulu's Cave, by which it is still known today.

The three stratified Middle Stone Age deposits were characterised by a progressive sequence of technologically more advanced industries. For example, in the lowermost stratum there was scarcely any trace of secondary trimming on the artefacts; this technical trait was rather more frequent in the second cultural stratum from the base, and more common and more elaborate on the artefacts in the third or highest stratum. An interesting feature was the presence of pieces of ochre and specularite in the upper two strata. Two of these took the form of deliberately fashioned specularite 'pencils', such as might have been used for making markings on a surface like human skin. The walls of the cave were made of quartzite and had a jagged and irregular surface that was totally unsuitable for mural paintings. Moreover, there was no trace of painting on any part of the cave's walls and roof. Body-adornment in life or on the dead at burial was the most likely explanation. Their evident interest in these colouring materials was corroborated when we found a well-shaped, neat little pestle with which it seems likely they had crushed pieces of ochre. What was so very interesting about the context of these finds is that these signs of artistic behaviour were manifest here as early as the Middle Stone Age. At that time, in 1947, nothing of the kind had been recovered in any cultural assemblages earlier than the Later Stone Age. Mwulu's Cave gave us the first signs of artistic activity at that great depth in time.

TWENTY YEARS LATER, in 1967, Adrian Boshier and Peter Beaumont, supported by Raymond Dart, found signs of purportedly early mining activity at Ngwenya in Swaziland. From a re-excavation of Border Cave, the scene of early modern-looking *Homo sapiens*, the presence of a notched baboon fibula, specimens of ochre, some of them ground, supported the evidence that artistic and by indirect inference symbolic behaviour had appeared already in the African Middle Stone Age. The discovery of red ochre slabs with man-made designs engraved upon them and dated to 78 000 years ago was recently announced by Chris Henshilwood. They were found in Blombos Cave in the Western Cape province.

Whether or not the evidence for early mining is accepted, and doubts have been expressed by some, all the other pieces of evidence, starting with the Mwulu's Cave pieces I published in 1949, add up to a strong case that, already in the Middle Stone Age and perhaps as long ago as 100 000 years before the present, human beings in southern Africa were capable of artistic and symbolic activity.

8 Hello to Sterkfontein

While I am recalling my early encounters with fossils at Makapansgat Limeworks and Mwulu's Cave, it is inevitable that my mind should turn to the Sterkfontein Caves. Although the bulk of my involvement at Sterkfontein occurred later in my career, I paid a brief visit to this area in 1945 on a student expedition, followed by many visits with my students and colleagues over the next 50 years.

The early history of the Sterkfontein and neighbouring caves is closely tied up with the quest for the main gold reef late in the nineteenth century. Two famous names in the history of the search for the main reef, Fred Struben and his brother Harry W. Struben, are reported to have purchased the two farms *Sterkfontein* and *Zwart Krantz* in January 1884. There, for seven months, they prospected diligently, seeking gold. We have no record whether they found any fossils.

The gold-mining industry had another impact on Sterkfontein. The Macarthur-Forrest cyanide process for the extraction of gold was introduced in 1891. It rapidly revolutionised the gold-mining industry and suddenly and dramatically increased the demand for lime. Lime prospecting and mining began at Sterkfontein and also at Taung, Makapansgat and a number of our other famous fossil-bearing dolomitic limestone caves. At Sterkfontein, an Englishman named Hans Paul Thomasset began quarrying for lime in or shortly before 1895. Guglielmo Martinaglia, a Piedmontese man whose father had helped to build the Simplon tunnel between Italy and Switzerland, entered on the scene with a lease on the Sterkfontein deposits. Working in close conjunction with Herbert Gladstone Nolan of Krugersdorp, Martinaglia blasted through into the underground caves, revealing to human eyes for the first time a vast network of caves containing a veritable fairyland of stalactite and stalagmite formations. It was probably late in 1896.

Immediately after Martinaglia had exposed the underground caves, and the news had broken in the press, there was an invasion of visitors. Unfortunately, much damage was done to some of the finest limestone formations by souvenir-hunters.

In the 1920s, the Glencairn Lime Company was excavating lime in the Sterkfontein area and George Barlow worked for this company for a time. When the Taung discovery was announced in 1925, Dart was sent some Sterkfontein fossils, including the skull of a large baboon, but these were not followed up. The reason is given in Dart's words: 'The Sterkfontein bones included the skull of a large baboon not greatly dissimilar from the living form, so I concluded that geologically this deposit must be relatively recent compared with Taungs [*sic*]'.[2] Interestingly – and with the irony of history – the most recent researches have shown that the relative dates of the two sites were in fact the other way round.

In 1936, one of Dart's research students, Trevor Jones, whose father had a farm at Doornspruit near Sterkfontein, visited the cave. In a letter to me written on 10 February 1984, from his home in Irene, Jones related:

> On a cold Saturday afternoon in July (1936) I walked past [R.M.] Cooper's hardware shop (in Krugersdorp) and saw the famous advertisement for bat guano. 'Buy bat guano from Sterkfontein and see the missing link.' On display was a heap of bat guano and some fossils. I introduced myself to Mr Cooper, who allowed me to take away the maxilla of a large cercopithecus, the lower jaw of an antelope and a crushed pelvis.
>
> On the following Monday I showed my specimens to [Sandy] Galloway and [Lawrie] Wells and started making plans to visit Sterkfontein. A girl friend with a car was most obliging and a week later we called at Sterkfontein. There we met Mr [George W.] Barlow standing behind a large table selling stalactites, stalagmites and fossils. Mr Barlow, who had worked at Taungs [*sic*] knew all about Prof. Dart and *Australopithecus*. He gave me about 43 specimens – mostly baboon skulls and baboon endocranial casts.
>
> I laid these specimens out on my laboratory bench and asked Dart, Galloway and Wells to examine them. Dart mentioned that Sterkfontein may yield another *Australopithecus*, but I know that he was too busy to

> follow this up. Later [Gerrit] Schepers and [William Harding] Le Riche came to see my collection. They, too, had visited Sterkfontein and had been given a baboon mandible by Mr Barlow.
>
> They immediately thereafter, drove over to report the find to Dr Broom at the Transvaal Museum. Broom wasted no time and the following day arrived to see Prof. Dart and the collection. Broom was most interested in the specimens especially the cercopithecus maxilla and the pelvis. We tried unsuccessfully to match the pelvis with the anthropoid pelves in the museum. Broom then left taking the pelvis and the cercopithecus maxilla with him . . .

The reason why I have quoted at length from Trevor Jones's letter to me is that, as he stated in his letter, 'My version of this visit has not been recorded fully.'

Jones's mention that Dart did not follow up his discovery echoes what Broom wrote in his letters some years later to Nicolette Wells, the wife of Lawrie Wells. These letters were kindly made available to me by the Wells's son, Gregory. When Nicolette Wells wrote to Broom after our 1945 and 1946 expeditions to Makapansgat, she told him that there were still hard feelings in the Anatomy Department that Broom in 1936 had 'taken over' Sterkfontein, which some of the research students, especially Trevor Jones and Gerrit Schepers, regarded as 'their site'. Ten years later there was anxiety in the department that Broom would likewise assume control of Makapansgat.

AS WE HAVE seen, Trevor Jones had obtained cercopithecid specimens from Barlow in 1936 and was working on them during his B.Sc. Honours year in the Wits Anatomy Department under Dart, Wells and Galloway. He intended that his fossil baboon project would blossom into a Master of Science dissertation.

In the same year, William Harding le Riche, another of Dart's science students, visited Sterkfontein and received from Barlow a baboon endocranial cast. Gerrit Schepers, a neuro-anatomist in the Anatomy Department, accompanied Le Riche on further visits during which they recovered a number of baboon crania, mandibles and endocasts. As Dart was away at the time and Broom had recently announced his discovery of a giant baboon (*Dinopithecus ingens*) in a cave near Hennops River in the Skurweberg, Le Riche and Schepers took the material to Broom. He accompanied them to Sterkfontein on Sunday 9 August 1936 and, eight days later, on 17 August, Barlow handed Broom fragments of South Africa's and the world's first *adult* australopithecine (ape-man), the first Sterkfontein 'missing link'.

Apparently both Jones and Schepers were irate at these developments. Jones was upset that the latest baboon finds from Sterkfontein had been taken to Broom: thereby, he said, Le Riche 'had mucked up his M.Sc. thesis which was to have been on the Sterkfontein fossils'.[3] The hard feelings in the Wits Anatomy Department towards Broom are portrayed in George Findlay's 1972 biography of *Dr. Robert Broom*:

> [Gerrit] Schepers felt that Broom minimised the importance of their [the students'] contribution, and after cooling their ardour took over the site for his own work. He echoed the sentiments of other staff members of the department: that Broom had seized the opportunity to take over 'their' site.[4]

Findlay defended Broom's actions in the following terms:

> Broom felt hurt at this view. In any case he could put the question of 'rights to the cave' very simply. Dart and his staff were not actually working at the site [Sterkfontein] in any real sense. The limestone was being burnt commercially all the time, and no arrangements had been made to preserve fossils disclosed during working operations. In fact, had Broom not gone out when he did, the Sterkfontein ape [*sic*] may not have been preserved. For this risk it seems strange to resent Broom's success . . .
>
> Dart's position was equivocal. It would have been nicest had his students been less disobedient and consulted him first, but who could guarantee that the discovery, which was to vindicate his reputation over the Taungs [*sic*] skull, would then have been made?[5]

Whatever animus there might have been between Dart and Broom in and soon after 1936 seems to have shortly evaporated. Dart appointed Broom as a part-time, visiting lecturer to the B.Sc. students in the Anatomy Department. Thus it was that the 79-year-old Broom lectured my B.Sc. class and led us to mount a brief expedition to Sterkfontein from 5 to 10 May 1945.

We scoured the Sterkfontein and Kromdraai Caves and included trips to two or three other sites. This visit was my debut at Sterkfontein. It was just nine years since Broom had recovered the first adult australopithecine there and six years before his death in April 1951. One of the side-trips we made from the underground circuit in the Sterkfontein Caves took us into a blind alley with breccia evident. This was the cul-de-sac to which I had the university later give the name Silberberg Grotto. Only two or three years before our visit, although we did not know it at the time, the art connoisseur Helmuth Silberberg had penetrated this grotto and (wholly without permit or permission) recovered a handful of fossils. One of these was a fine specimen of the lower jaw of a very primitive hyena. When Abbé Breuil, the doyen of French archaeologists in the first half of the twentieth century, saw the specimen in Silberberg's gallery in Johannesburg, he obtained permission to take it across to Broom at the museum. Broom recognised it as a species of extinct hyena that in other areas had proved to be of high antiquity.

Broom was excited by Silberberg's find because of its presumed great age. Up until that stage, Broom had considered the Sterkfontein deposit to be too recent for its hominid contents to have any bearing on the origin and evolution of mankind. When the new specimen of old hyena came into his hands, he inferred that Sterkfontein was much older than he had thought. He now came to believe that the Sterkfontein ape-man was so ancient that it could have had some bearing on human origins. It would therefore be worthwhile for him to resume excavations at Sterkfontein.

Consequently, after the war-time interruption of field activities, Broom went back to restart work at Sterkfontein in 1946.

At that point the Historical Monuments Commission, the body which gave permits for excavations in those days, refused to give Broom a permit to dig at Sterkfontein. They held that he was not paying sufficient attention to the stratigraphy (layering) and geology of the cave deposit. Broom was indignant. He moved to Kromdraai just over a kilometre away and started excavating there. The Monuments Commission gave him a permit to dig at Kromdraai but not at Sterkfontein. Defiantly, Broom moved back to continue digging at Sterkfontein.

The perverse streak in Broom's temperament led to one of the most famous and important discoveries of a fossil hominid. On 18 April 1947, Broom with his assistants John Robinson and Daniel Mosehle, blasted out the complete cranium of what Broom called *Plesianthropus* (hence 'Mrs Ples'). Was it poetic justice that Broom made this discovery of what he called 'the most valuable specimen ever discovered' while he was without a permit?

Broom's reaction to the furore that erupted was bitter and cynical:

> It is in my opinion the most important fossil skull ever found in the world's history. The discovery created a considerable sensation in both Europe and America. In South Africa, however, the principal reaction was of a different nature ... We had broken the law. We had flouted the Historical Monuments Commission; and the fat was in the fire. The Commission held a full meeting at which I was unanimously condemned.[6]

Broom was forced to stop his work at Sterkfontein. However, by late June 1947, a referee appointed by the Historical Monuments Commission, Professor Lombaard (a geologist at Pretoria university) had submitted his first report on Broom's Sterkfontein site in which he stated: 'I am satisfied that there is no immediate danger of destroying evidence based upon geological stratification ...' Broom was duly awarded a permit and his campaign of wresting as many ape-men as possible from the Sterkfontein deposits was legally able to continue.

Much of my insight into the details of these events has only emerged in subsequent years. However, as a student I was fascinated by the personalities involved in these manoeuvrings as well as by the politics (and science) of the various discoveries being made. My education in all spheres of life was an ongoing and deeply exciting process.

9 | Political awakening

My immediate family was largely apolitical, which meant they did not talk much about politics and they did not play any active part in the political contest between the two leading white parties of the time, the Nationalists and the United Party. In addition, education in government (public) schools in South Africa was segregated long before apartheid was formalised. Thus it was that I came to Wits in 1943 as an innocent abroad – I came from a politically naïve background and had attended all-white primary and secondary schools.

I immediately had my first lesson in race relations. On the campus and in my medical class, there were students of various colours and ethnicities. Suddenly, in the fresh air of Wits, a university that admitted students irrespective of their race, I found myself sharing desks, benches, laboratories, dissection halls, auditoria, libraries and the University Great Hall with students of diverse ethnic backgrounds. It was the finest lesson a young idealist like myself could have had. Very quickly I became an ardent supporter of 'non-racialism'. Without the need for any political speeches or sermons, the non-racial composition and policy of Wits became part of my way of life, my spirit and my ethos. Black students and their white fellow students shared the same problems and difficulties, both in the academic arena and on the tennis courts.

Then a delegation of church representatives visited the university principal, Humphrey Rivaz Raikes, and complained about 'mixed sport' on the campus. Nervously, the university felt obliged to put a stop to the practice and, amongst other activities, my weekly tennis games with Abdul Hak Bismillah and other fellow students came to an end.

The official practice of the university was declared to be 'academic non-segregation, but social segregation'. This was a dreadful step backwards and was a faint-hearted surrender by the University Council three or four years *before* the National Party government came to power on a platform of avowedly segregated education at all levels.

MY EARLIEST LINK with Jan Hendrik Hofmeyr, that supreme exponent of non-racialism, was at my first graduation ceremony, on 16 March 1946, at which he presided. It was the first time that I nervously made my way across the huge stage to receive a degree (my Medical B.Sc.). In those days, Wits graduation ceremonies took place in the Johannesburg City Hall and were noisy and raucous occasions as a large percentage of the student body was present to sing and cheer their favourites across the stage.

On this occasion, Hofmeyr – who was the university chancellor as well as the Minister of Education, Minister of Finance and deputy prime minister – delivered an address that some regarded as the greatest and most courageous oration of his career. He reminded his audience of the four freedoms of the Atlantic Charter: the freedom of speech and expression, freedom of religious worship, the freedom from want, and the freedom from fear. To these four freedoms, Hofmeyr proposed that a fifth freedom be added: freedom from prejudice. As he pointed out, 'The chief cause of our large measure of failure to realise the idea of freedom in our land is the strength of prejudice – race prejudice and colour prejudice.'

To us young neophytes he commended this fifth freedom as worthy that we should fight for it. 'I shall put it more strongly than that,' he added. 'May you be prepared to say with Thomas Jefferson, "I have sworn upon the altar of God eternal hostility against every form of tyranny over the mind of man" – and here in South Africa the greatest evil of all is the tyranny of prejudice.'

Hormeyr's advice is no less cogent today, 60 years later. The new South Africa has changed the constitutional structures under which it is governed; it has in place one of the world's most enlightened constitutions and is changing the socio-economic dispensation, but such changes are not going to sweep away the deep-seated racial prejudices in many people's minds. Freedom from prejudice – all kinds of prejudice – is still an essential goal.

THE NATIONAL UNION of South African Students (Nusas) was founded by Leo Marquard in 1924, on the model of the National Union of Students (NUS) in England. At an historic meeting in Bloemfontein in the winter of

1945, Nusas resolved to admit Fort Hare, the South African 'Native College' to membership. This ended a long period in which Nusas had sat on the fence, hoping first to restore unity between English and Afrikaans-speaking white students and then to admit black students. (The Afrikaans universities had left the National Union in 1933–36 to form their own unilingual, uniracial, unicultural organisation.) From that moment in 1945, Nusas was open to all students, irrespective of race, at all universities in South Africa.

I and many of my fellow students in the Faculty of Medicine at Wits rejoiced at this development. I decided that I did not want to remain simply a passive member of Nusas; rather I wished to play an active role. It so happened that the national president of Nusas was a senior medical student at Wits. Taking my courage in both hands, I approached him and said that I should like to play an active part in Nusas. How should I set about it? He looked at me and said, 'Nusas doesn't need your kind of person.' I was stung. It could only have been because he had rightly surmised that I was Jewish. I resisted the urge to punch him on the nose: that wasn't my style. But his disdainful rebuff goaded me to my goal with even greater resolution than before. I went on to become a national office-bearer for five years, three of them as national president and to be dubbed by an historian as 'the chief architect of the post-war Nusas'.

A MAJOR PLANK in the National Party's electioneering in 1948 was to compel the universities to segregate. There were threats that so-called 'white universities' (an epithet which Wits, as well as the universities of Cape Town, Rhodes and Natal, had never accepted) would be forced to exclude students of colour. Staff members (or 'faculty' in the American sense) would likewise be limited to whites at 'white universities'. Black students, whether African, Indian or Coloured, would have to attend the 'racially appropriate' institutions.

Just two months after the National Party had come to power in the general election of May 1948, I was elected as the president of Nusas at the annual congress. The fortunes of the National Union were at a low ebb: three of the leading centres, Cape Town, Durban and Rhodes, were under notice of disaffiliation, and it looked as though I had inherited a sinking ship.

As I researched through the Nusas files, I discovered that Jan Hendrik Hofmeyr had been elected honorary president of the National Union some fifteen years before. This appointment had never been terminated or superseded. I concluded that he still held the position. Accordingly, I called on Hofmeyr in parliament in Cape Town and explained the gloomy situation in which, as its new president, I found Nusas. He responded, 'If you had come to me three years earlier, I might have advised you to defer a decision on Fort Hare and to try a little longer to persuade the Afrikaans universities to rejoin Nusas. But now that the decision has been taken, there can be no going back. If the Nusas ship sinks as a result of the 1945 decision, then it is your duty as captain to ensure that it goes down with all its flags flying.'

I did not like the first part of Hofmeyr's remarks – it was precisely the admission of Fort Hare that had first aroused my serious interest in Nusas. However, the advice he gave me became a guiding star in my actions over the next three years. Happily I did manage to keep the flags flying and the ship afloat. Nusas was reunited; many more centres were brought in, and it was welded into a powerful organisation that played an important part in the fight against apartheid.

My third and last link with Hofmeyr came four months after our conversation. He died in Pretoria on 3 December 1948, at 54 years of age. Two days later I had the sombre duty to represent the students of South Africa at the funeral of the man whom many had regarded as South Africa's great liberal hope.

SHORTLY AFTER MY election as Nusas president in 1948, the executive committee and I launched the first anti-apartheid campaign in the South African universities. We circularised well-documented statements to vice-chancellors and presidents of international universities all over the world, stressing two principal facets of the government's policy that we and many others found objectionable. Firstly, restrictions were to be meted out to black students that were not applied to white students: this was sheer racial discrimination, we pointed out. Secondly, we deplored the inroads into academic freedom that were being threatened.

My first circular on these matters was sent to all university staff members in South Africa, members of parliament and the provincial councils, cabinet ministers, ministers of religion, student leaders, whether or not they belonged to Nusas, public figures and, of course, to various individuals and organisations around the world.

The South African response was mixed. A number of prominent figures wrote strongly supporting our stand, but a few were not so complimentary. An anonymous academic at Wits sent my carefully worded document back to me with the word 'BALLS' scrawled over each page. At the end, the person added, 'Why don't you marry a kaffer woman, Tobias? That's all you're good for!' (The word 'kaffer' in the South African environment is a highly derogatory term for black African people.)

From overseas we had wonderful support from such luminaries as the vice-chancellors of Oxford and Cambridge, and the presidents of Harvard and Yale. Some of those who had received the letter wrote to the prime minister and the Minister of Education, and to vice-chancellors or rectors of South African universities. The vigour of the international stir gave the government pause in its avowed intention of legislating apartheid into the universities. I am convinced that our campaign and the reaction of overseas academics, not to mention of student leaders from many countries, played a major part in compelling the government to delay implementing apartheid in the universities of South Africa. Although we were expecting the drop of the guillotine every year, month, week, it was not until 1959 that the legislation in question was tabled under the euphemistic title, 'The Extension of University Education Bill' (see Chapter 26). I believe the student leaders of South Africa, at first under my stewardship from 1948 to 1951, and then under the Nusas leaders and Students' Representative Councils that maintained and intensified the struggle, played a fundamental role in staving off the legislation. They did so even in the face of security police wielding tear-gas as they invaded the Wits campus, and spies being planted at the university and in the Students' Representative Council itself.

Although our campaign kept legislated apartheid out of the universities for close on a dozen years, there were many administrative steps the government could and did take under existing legislation.

SOON AFTER THE National Party came into office, the government scholarships for African medical students to attend Wits were summarily stopped.

These scholarships had an interesting history. They were initiated in 1941 during the Second World War when black South Africans (who had usually travelled to the United Kingdom for their clinical years of study) were unable to travel overseas because of the blockade and counter-blockade on the high seas. Accordingly, the South African government created a series of medical scholarships for five African students a year to attend Wits. These medical scholarships were renewed annually.

It was a paradoxical situation: the Union government was sponsoring the medical training of black students at Wits, in contrast with the inhibition of such developments in the ensuing political phase. It was these scholarships that the nationalists terminated soon after they came to power. The government's objective was to further the apartheid policy by cutting these black students off from admission to what government chose to call a 'white university'. This step was an administrative measure for which no new parliamentary enactment was necessary.

Nusas, led by me, exposed this perfidious action for what it was and took steps to replace these government scholarships with student-sponsored awards. In a crowded mass meeting in the Great Hall, Wits students resolved to set up the African Medical Scholarships Trust Fund. Most students willingly agreed to add the sum of two pounds and ten shillings on to their fees each year. With thousands of students participating, we were soon able, with the co-operation of the university, to replace some of the discontinued government scholarships. Through Nusas we widened the appeal and received support from students at other South African universities as well as many overseas contributions. This inspired response from various sources enabled us to replace all of the government scholarships and even to increase the number awarded each year. We successfully maintained this until 1959 when the government declared the scholarships illegal.

The end of the government medical scholarships was only one of a series of administrative blows against the Open Universities. At that time, South African Asians (Indians) who wished to cross from one province to another were obliged to obtain an interprovincial permit. The government refused these permits to several Indian students from the province of Natal who had been accepted for enrolment at Wits (which was in another province, then called Transvaal). The law permitted the Minister to give no reasons for his decisions, despite our protests and requests for reversal of his refusals. Such a seemingly minor thing might have been overlooked as a trifle, but for the fact that we exposed it publicly as yet another governmental action to thin the ranks of students of colour at Wits University.

Another seemingly arbitrary decision was the refusal by the Minister to renew the visa of a Mozambican student, Eduardo Chivambo Mondlane. He was studying in Johannesburg on a scholarship from the Christian Council of Mozambique and was only six months from completing his degree course. His visa to study in South Africa was due for renewal but this was refused by the South African government. Mondlane had to return to Mozambique. He had not been 'dismissed from the university for being a "foreign native"' as was later mistakenly asserted in accounts of Mozambique's history by an American professor, Thomas H. Henriksen. Eduardo was instead another victim of the campaign of administrative exclusionary measures by the government.

As soon as the student body received word of Mondlane's virtual banishment, we called a mass protest meeting of students at Wits. We held what I believe was the first open-air meeting in the amphitheatre next to the swimming pool. The meeting was chaired by the president of the Students' Representative Council, Sydney Brenner (later a Nobel Laureate in 2002), whilst I moved the motion from the floor. It expressed our deep shock at the refusal of Eduardo Mondlane's visa and earnestly requested the Minister to reverse his decision. According to the report in the *Rand Daily Mail* the following morning, the resolution was passed by 700 for, to 4 against.

When Brenner declared the meeting closed, a figure detached himself from the back of the crowd and came towards me, asking for a copy of the resolution. It was immediately apparent that he was probably a security officer in plain clothes. I said in a loud voice so that students around me could hear, 'You'll have to ask the chairman of the meeting.' And I strode swiftly down to Brenner. The man followed me at a little distance. When I reached Brenner, I said out of the corner of my mouth, 'The cops', a phrase I do not think I had used before nor have I since. Brenner confronted the man and asked him what he wanted. Eventually he confessed that 'my minister wants to know what you b—s are getting up to at university'. We handed over a copy of the resolution – after all, it was public property now. It was my first brush with the security police, though certainly not my last.

Our request to the Minister to reverse his decision was, not surprisingly, ignored. Fortunately, Ieuan Glyn Thomas, the university registrar (and later vice-principal), arranged for the lecture notes for the ensuing six months to be sent to Mondlane to enable him to continue his studies from afar. Moreover, when the time for the finals arrived, a Wits official took a set of the exam papers to Mozambique and Mondlane wrote his exams under invigilation. He passed and his degree was conferred *in absentia*. 'As long as I live', said Mondlane, 'after the way the Witwatersrand University has treated me, I can never be regarded as anti-white.'

Mondlane later became the leader of Frelimo, the movement for the liberation of Mozambique from Portuguese rule. He lived in Dar-es-Salaam and it was there, at a friend's house, that he received a parcel post-marked from Moscow. When he opened it the parcel bomb exploded. It was 3 February 1969 and Mondlane was not yet 50 years old. The perpetrators have never been uncovered.

IN MY PUBLIC protests, as president of Nusas, against the government's exclusionary actions I was at pains to show that each incident was not an isolated episode. It was part of an operation to frustrate and inhibit Wits University's open-door admissions policy. The government's actions constituted an insidious pattern: to cut down the number of black students at the Open Universities until there were so few left that legislating apartheid into the university environment would seem to be a small and negligible step, a mere formality.

It was a truly dark time, both for me personally and for the country as a whole, when the university apartheid legislation was finally tabled and passed by parliament in 1959.

10 | African eye-opener

Shortly after I received my medical degree in 1950, I was mulling over what direction to follow. A compulsory internship had been introduced for all South African medical graduates a year or two earlier. If I wished to become a registered medical practitioner in South Africa, I would have to complete the one-year internship. I was not at all sure I wanted to become a practising doctor.

In this state of indecision, I was invited to Raymond Dart's office. He got down to business right away. A vacancy on his staff had recently come up with Lawrence Wells's move to the Edinburgh Anatomy Department. Would I take up a position as a lecturer in the Wits Department of Anatomy? It was an offer I could not decline. I had already spent the best part of six years attached to the department, as a research student and part-time teacher, and my commitment to it was fervent.

I settled in to my new job and almost immediately became heavily involved in teaching students of medicine, dentistry, physiotherapy, occupational therapy and 'science'. Towards the end of June 1951, I received a summons to Professor Dart's office. He said: 'The South African government has approached the university to second a suitably qualified person to a French expedition to the Kalahari desert. One of its aims is to study the Bushmen in the Kalahari. I propose that you join the French Panhard-Capricorn expedition as anthropologist and take some measurements on them.'

'I haven't had enough experience to take on a responsibility of this order,' I expostulated. Dart felt it would be a unique opportunity to give me a wonderful fund of experience and, in the end, he proved right and my fears ill-founded.

When I complained that I had not had enough tuition in anthropometry (the measurement of the human body), Dart's peremptory reply was, 'Well get out there and learn how to do it!' I did not at the time find this advice very helpful, especially as the expedition was due to depart barely six weeks later.

This expedition, led by the French explorer and geographer Francois Balsan, was to travel across Africa, more or less at the level of the Tropic of Capricorn. It ushered me into the world of the San (or Bushmen), and the living peoples of Africa nearly became my main anthropological thrust. I spent the next twenty or more years involved in various programmes of research on the Bushmen.

In the weeks before the expedition's departure I frenetically taught myself anthropometry, acquired instruments that were affordable, and schooled myself in other human biological variables, including genetical ones. I did not believe that a mere set of bodily measurements was adequate for a field study of these rather rare survivors of one of the world's few, still existing, hunting and gathering peoples. I prepared data sheets on which to record scores of observations on each individual.

As a recently qualified medical man, I was also expected to be the expedition's physician. I built up a medicine kit that was far more than a mere first aid outfit. I was equipped to handle even a case of appendicitis. In the event, the only surgical procedure I was called on to do was on the first night of the expedition. It was a 'cut-down' over the scalp of Francois Balsan, to remove a sizeable thorn of a camel-thorn tree. This had entered the scalp, broken off and rotated sideways into a deep scalp layer. With light from a lamp and a few candles, I performed the minor operation. The thorn was extracted cleanly and my patient made a good and uneventful recovery. For the remaining three months of the expedition, I was fortunate to have no occasion to open my surgical kit, save for a splint, crepe bandages and dressings.

BY THE TIME August arrived, I was fairly well equipped, materially and scientifically, for my 'venture into the interior'. At 25, I was one of the younger members of the expedition team. We were given a grand send-off party at the offices of Union French Industries in Germiston, close to Johannesburg. Among the fairly large crowd present were Colonel Tainton and his wife Doreen, who had just arrived back from their search for 'The Lost City' of the Kalahari. Weeks later, at one of our overnight camps in the Nosob River, we found a French brandy bottle hanging from a camel-thorn tree by an item of Doreen's elegant lingerie, with an affectionate note to Francois Balsan and words of encouragement appended.

The story of the venture, known as *L'Expédition Panhard-Capricorne*, has been told by its leader, Francois Balsan, in his book of the same name published in 1952. Over the next eleven weeks we travelled through four southern African countries: from South Africa to Botswana to Namibia to Botswana to South Africa, to Mozambique and back to South Africa. My personal journeying involved some 6 800 kilometres. It was the first of my overland travels out of South Africa and it was to be the first of many.

The enchantment, the lure, of our continent was borne in upon me and the impressions it left are still with me today. The crossing from ocean to ocean, more or less along the line of the Tropic of Capricorn, showed me a formidable array of ecologies, from semi-desert to forest, from fossil rivers to flowing rivers, river banks, lake shores and seashores. I encountered teeming humanity, from the light, yellow-brown San to the taller, dark-skinned black Africans, from hunters and gatherers to pastoralists, agriculturalists and urbanites. The diversity of languages was richly apparent, even to someone like me who does not claim to be a master of African languages. The material culture of the living populations was floridly prolific and, in many respects, showed me how closely the culture was adapted to the terrain. The life-ways even formed a major basis of the adaptability of the peoples to their environments, even more, it seemed to me, than did their genetic make-up.

ON THE PANHARD-CAPRICORN expedition I was plagued by numerous and serious asthma attacks. My asthma had been diagnosed for the first time while I was on a Nusas visit to Grahamstown three years earlier. I was accompanied by Sydney Brenner. We went to stave off an attempt by the Students' Representative Council and the student body of Rhodes University to disaffiliate from Nusas. This came in the wake of the National Union's assuming a non-racial status and adopting our proposal for an anti-apartheid policy.

We stayed in a small private hotel with minimal facilities. The air was charged with very high humidity and periodic heavy rain. During the night, I found it more and more difficult to breathe. We had not reached such bronchial ailments as asthma by that stage of our medical studies at Wits and I did not know what was the matter with me. As the attack grew more severe and my chest ever tighter, a feeling of impending doom came over me (as described in clinical textbooks of the time). The guest-house telephone was disconnected from about 6.00 p.m. I dragged myself out of bed, put on my raincoat and shoes and, armed with an umbrella, ventured out in the pouring rain. Wheezing more and more, my breathing increasingly laboured, I walked right around the large central town square of Grahamstown, looking for the tell-tale brass plate of a medical practitioner. I had turned left from my lodging and circumnavigated the square clockwise. Feeling by now very poorly and having all but completed the circuit, I was returning to the hotel on its opposite side – and there, alongside my lodging – but on its right – was a doctor's brass plate! It was about midnight when I rang the doorbell. Eventually a gaunt Dickensian figure opened the door, the doctor himself clad in a nightshirt and wearing a tapering sleeping cap. He took one look at me and said, 'How long have you been an asthmatic?' The diagnosis was made. Then he plunged a needle into my arm and the adrenaline rapidly brought relief.

When I set out for the Kalahari three years later I did not know about all my allergies, which David Ordman was later to decipher at the South African Institute for Medical Research. My most severe sensitivity was to equids, the horse family, to which, of course, belonged donkeys. Dog dander was my second-to-worst allergen. Other animal skins, like jackal-skin karosses, were also allergenic to me, but curiously not cats or other members of the cat family.

As I suffered repeated and severe asthma attacks in the Kalahari I was apt to blame these episodes on the dust and desert sand, little realising that the frequent donkey-cart rides that I undertook for various purposes and the karosses we sometimes sheltered under were the sources of my difficulties. The only suitable medication I had with me was ephedrine tablets (the modern inhalers and medicines used to treat asthma were decades into the future) – and there was a limit to the amount of ephedrine I could take by mouth.

IT HAD ALWAYS been one of Francois Balsan's objectives on the expedition to seek the ruins of the supposed Lost City of the Kalahari. This legend goes back to 'Farini the Great' who in 1885 travelled with his son and a certain Gert Louw in the Kalahari. The name Gilarmi A. Farini was a pseudonym. His real name was William Leonard Hunt and he had been born in New York in 1839. Farini's 1886 book, *Through the Kalahari Desert*, described his discovery of the ruins of a 'Lost City' near Kij Kij (or Ky Ky) on the Nosob River in the south-west of Botswana. The area in question would have been in the Kgalagadi Transfrontier Park, which has recently been brought into being and straddles the frontier between South Africa and Botswana. Farini's announcement caused excitement and over the ensuing years many expeditions were mounted in search of the fabled Lost City.

Balsan had made an earlier excursion to the Kalahari in 1948. Carefully scrutinising Farini's account and sketches, Balsan calculated the approximate position of Farini's 'discovery'. Before the Panhard-Capricorn expedition set out, Balsan chartered an aircraft and carried out an aerial survey of the area in question. No sign of a Lost City was detected.

Our ground survey began when we reached the Nosob River and Le Riche's camp. Back and forth we went around about Rooi Pits and Ooikolk. Beautiful sections were exposed, featuring red sands, at the top, the base of which was some 8.5 metres above the river-bed. Between these two extremes were a pebble-bed and two terraced horizons with a calcrete bed intercalated. Further north, an even thicker layer of limestone was apparent. People of the Middle Stone Age and Later Stone Age had lived on, or in the lee of, the dunes. More 'architectural' forms seem to have been sculpted from the limestone, including two concentric circles of calcrete pieces, looking suggestively human in origin. Overshadowing these probable products of natural weathering was a magnificent pebble-bed with freshly struck artefacts on pebbles and derived archaeological remains of sandstone.

By now I began to think that, rather than concocting the story of the Lost City, Farini had misinterpreted natural calcrete formations and natural weathering forms along the Nosob River. I was even able to show Balsan some of these natural formations that resembled Farini's illustrations in his book. Balsan and I were now convinced that this was the explanation behind the tale of the Lost City.

By the time we laid the Lost City to rest, I was thoroughly disgruntled. Twenty-three days after we had left Germiston, I had not yet been able to study any Bushmen, the whole reason I had been seconded to the expedition. True, I had been able to familiarise myself with something of the prehistory of the Kalahari dwellers. I had identified more than a dozen previously unreported archaeological sites and had made a number of carefully labelled collections of artefacts. True, at intervals from the beginning, I had been in contact with isolated Khoe-Khoe persons, and I had examined a few Khoe-San skulls in the Windhoek Museum. Nevertheless, I was anxious to turn my attention from archaeological reconnaissance to meeting and studying Bushman groups. Fortunately, over the next few weeks I was able to gather enough data to satisfy what I felt was the true purpose of my presence on the expedition.

ONE OF THE sequels to the Panhard-Capricorn expedition was that I published a number of articles on the San, especially from a human biological point of view. My results, especially the evidence that large numbers of San were still alive and that culturally they were extremely well adapted to life in the desert, aroused great interest at Wits University, in South Africa in general, and overseas, especially in England and the United States.

As a gesture to the French party led by Francois Balsan, I published the largest of my manuscripts on the San in French, as a series of four parts in the well-known French periodical, *L'Anthropologie*. A great deal of my original data taken from the San on the French expedition appeared in these articles. If I reread some of my conclusions from those studies, I realise now that I was under the influence of Raymond Dart's typological approach to the analysis of the 'racial affinities' of African peoples. It was an influence that I was soon to shake off, for it was an approach that was totally at variance with the lessons of human genetics. Moreover, typological thinking goes hand in hand with stereotyping and forms the basis of what the Germans used to call *Rassenkunde*. Surely such thinking was to be eschewed in an age when we were learning so much more about how the hereditary material worked, an era when we had scaled new heights in the understanding of the structure of DNA and of how genetic information was encoded.

In most of my writings on the San flowing out of the Panhard-Capricorn expedition, I was concerned with such issues as the numbers of surviving Bushmen; how to determine the degree to which their environmental adaptation depended on their genes and on their culture; the first signs that, under better living conditions – where government was digging boreholes and distributing famine relief – the San were becoming taller in adulthood than their forbears of the same clan had been (a trend that we call the positive secular trend); as well as the concept of race and limiting factors in the determination of the South African races. This was a rebuttal of the apartheid government's Race Classification Act: apart from the unethical attempt to force every person to be labelled racially, I tried to show that what the government was attempting to do was unscientific.

Another important consequence of the expedition and my reports was the decision of Wits University together with the London-based Nuffield Foundation, to set up the Kalahari Research Committee under my chairmanship. Over the next fifteen years, from 1956 to 1971, this committee was to mount many-sided expeditions to study especially the San. When I look back at what was 'received wisdom' about the San in 1951 when I started my researches, I cannot but marvel at how many myths have been overturned by the studies of my colleagues and myself in the 50 years since my first encounter with these remarkable human beings.

Another profound lesson I gleaned from the expedition was the prehistoric wealth of Botswana and the adjacent territories. The semi-desert of the twentieth century had been preceded by times of great abundance, which allowed human populations to be fruitful and to multiply, from Windhoek to Pafuri and beyond.

Above all, the Panhard-Capricorn expedition taught me how important it was to keep my eyes wide open: well-trained powers of observation are the most important equipment required for success in fieldwork.

11 The happiest year of my life

My Ph.D. thesis on the chromosomes and sex-cells of the gerbil nearly went up in smoke – literally – shortly before I submitted it in 1953. A person who shall remain nameless inadvertently left a hot iron on and placed it on the pile of hundreds of typed pages. The iron burned its way down through a hundred or more pages, many of which I had no copies of, before the smouldering stack was discovered and rescued.

During the examining process, as is customary, I did not know the identity of my examiners (one internal and two external). I learned afterwards that, while two of my examiners had reported timeously (and favourably) to the Higher Degrees Committee, the report of my third examiner, who was Theodosius Dobzhansky (see Chapter 30), was delayed. The university sent him a telegram. His response came swiftly by way of a return telegram that intimated unequivocally and enthusiastically that 'the thesis of Mr Tobias was the most outstanding and thorough genetic study of a mammal he had ever read, and that there should be no hesitation in awarding him the degree' – or words to that effect. Considering who had sent the cablegram and the positivity of his judgement, the university, for the first time in its history (I was later told), had no hesitation in accepting that telegraphed report in lieu of the usual longer and more ponderous analysis. My Doctor of Philosophy degree was conferred on me in 1953.

From the Nuffield Foundation of London, I was fortunate enough to be awarded a Nuffield Senior Travelling Fellowship to spend a post-doctoral year in England. I chose to go to Cambridge because my training under Dart and Wells at Wits had been almost devoid of a mathematical and a statistical approach to anthropology.

I had realised that I could not hope to make any contribution to science without the use of mathematics and statistics, even if these subjects were difficult for me. The Biometric School had been vigorously active in

England and people like Egon S. Pearson, Francis Galton, Hans Kalmus, Geoffrey Mourant and Jack Trevor had carried the message forward. Although I was not wedded to the Biometric School's approach, I felt it was necessary to expose myself to their methods to a greater degree than I could hope to learn in the Wits Anatomy Department at that time. At Cambridge, in the Faculty of Archaeology and Anthropology, was a physical anthropology centre called the Duckworth Laboratory. It was directed by Jack Trevor and I spent the best part of 1955 working in the Duckworth.

It was the happiest year of my life.

Cambridgeshire is a low-lying county and much of it is comprised of The Fens. It extends as flat, marshy country eastwards to the sea. The air is very humid and it has a reputation for being unkind to asthmatics; indeed, it is said to be the worst county in England for asthma. When I arrived in England, it was New Year's Eve, 31 December 1954. The boat-train took other passengers and me from Southampton to London. Along with tens of thousands of others, I greeted the New Year in Piccadilly Circus. Thousands of voices were lifted in song as a light pattering of snowflakes added an extra zest to the scene. A day later I took the train up to Cambridge and moved into temporary lodgings near the station. Already my chest was heavy with broncho-spasm and over the next ten days or so I had difficulty tramping the streets of Cambridge in the grey, misty cold to find more permanent accommodation.

On my first Sunday afternoon, I was sitting dispiritedly at the window watching the leaden skies and the light snow. Suddenly I saw a man trudging towards the station and recognised him as none other than a friend from Wits. Seeing him was like a lifebuoy thrown to a drowning soul. I rushed out into the cold and shouted his name. It was Percy Barkham. He had been a member of the science class with Syd Brenner, Leo Bagg, Gloria Gearing and others. When I told him of my misery, he kindly changed to a later train back to London and we went out to dinner in one of Cambridge's innumerable restaurants.

I finally moved into digs in Jesus Lane, opposite Sidney Sussex College. The lodgings were run by a Cambridge lady married to Stan Maletka, a Polish immigrant. From there each day, it was a fair walk down the Petty Cury, across the Market Square, into Downing Street – and so into the Faculty of Archaeology and Anthropology.

It took some time to become attuned to the ambience and the people of Cambridge, but I slowly made a circle of friends. The librarian of the Haddon Library, Mary Thatcher, was a lovable personality. She and I took an immediate liking to each other. She helped me adapt to the ways of Cambridge and gradually I found myself enjoying it more and more. One day I went into the Haddon Library and asked Mary for the 1921 edition of Martin's *Lehrbuch der Anthropologie*. I was told, 'We do not possess it' – and then, with an airy wave of her little arm around the library, she said, 'We do have other books though'. She and I tittered over this typically Cantabrigian riposte for months afterwards.

One day in April, she said, 'Phillip, do go down to The Backs and have a look at some little heaps of snow, then brush away the top of a heap and you'll have a surprise.' I did as she suggested and there under the majestic oaks were a number of small heaps of light crumbly snow, rising from the surrounding flat layer of snow. I brushed away the summits of two or three and, there, within each, was a tiny, ice-cold snowdrop, white and alive, pushing up the overburden of snow. Some of the heaps were being pressed up by golden aconites. It was a magical moment of revelation: spring had arrived!

My long-dormant poetic instinct came alive. I remembered with new meaning Percy Bysshe Shelley's words: 'If winter comes, can spring be far behind?' And Robert Browning's, 'O to be in England now that April's there'. I exulted and my mind, just like my lungs, filled afresh with good, clean air. That moment, for which I have Mary Thatcher to thank, was a turning-point in my year-long stay at Cambridge. For once a real, positive sentiment of happiness flooded over my being: happiness in the present tense!

I learned to punt on the Cam and a favourite Sunday afternoon pastime was to punt the few kilometres to Grantchester for high tea in the orchard. Another poetic allusion came to mind, 'Oh! yet stands the church clock at ten to three?/And is there honey still for tea?' The romantic swelled in me as the punt was moored on the bank leading up to the orchard. There was the church clock and, believe me, it still stood at ten to three, as it had done for Rupert Brooke in 1912, and for aught I know stands there still at the Old Vicarage. And there *was* honey still for tea, as well as all the other delicacies that distinguish high tea in the orchard from mere afternoon tea.

During my year at Cambridge I drove about once a week or fortnight to London, to work at the Natural History Museum in South Kensington. I used the evenings to attend plays and concerts. One of these visits gave me the chance to hear a memorable recital by Arthur Rubinstein in the Royal Festival Hall, early in June 1955. The audience gave him an American-style ovation and he was prevailed on to give five encores, the last with hundreds of people packing up close, on and around the stage.

After the concert I met up with Saul Zwi, my former research student and friend who was working at Hammersmith Hospital, his wife, Helga, and other mutual friends. As a professor of respiratory health at Wits, Saul was later to be my asthma specialist.

I arrived back at Cambridge at 3.25 a.m. As I still felt wide awake, I sallied on to The Backs along the Huntingdon road, parked my car at the entrance to Clare College, and walked alone in the dark, grey dawn. Only the rooks were astir, cawing with a vengeance in their rookery at King's; but as a fresh morning breeze stirred the trees and the 'giant lichens of the sun' crept up the eastern sky, the more melodious chirrup of another bird was heard, then another, and another, until all the world was filled with the enchantment of the dawn chorus. White and ghost-like in the greyish light, a swan glided towards me in one of the many affluents of the Cam. Later, when I caught it off guard, I saw its gracefulness and dignity laid aside, as it had an early morning splash and spread its wings with a great flap-flapping. On a wooden rail just in front of me, a blackbird sang lustily, if a little 'throatily', unaware of my presence. The breeze freshened, the light brightened and I finally repaired to my room and the relative comfort of my hollowed bed, having placed a scrawled notice on the door: 'Do not wake for breakfast – PLEASE!!' It was 5.30 a.m.

Whit Monday, 1955, was our second successive day of glorious hot sunshine. Victor Pollak of Wits, who was doing renal research in Chicago, spent the weekend as my guest. Early in the morning, when boats were still available, we took a punt and I started punting upstream towards Grantchester. I managed very nicely, did not fall off (contrary to what is said to have happened the first time Syd Brenner went punting on the Isis!) and took us about a kilometre upstream. When Vic took over, the fun really started – our craft described large, lazy circles, lay athwart the current, colliding with perfectly respectable, oncoming punters, headed violently for the mud on one bank or jerkily for the reeds on the other – or moved slowly backwards! Eventually, we had a delightful picnic on the grassy banks of the river near Grantchester, complete with buttercups, daisies and pale ale.

'The Mill' is a famous little pub at the bottom of Mill Lane, where draft Merrydown cider is the order of the day. Everyone takes his or her drink out on to The Backs, the bridge or the grassy slopes and enjoys the open air on warm summer evenings. Hundreds of young men and ladies, with refined Cambridge accents, can be seen there on some evenings. Occasionally there is a loud splash as someone goes fully clad into the water – but whether pushed or dared or simply to escape an approaching proctor, it is difficult to say. The Mill is considered by some to be one of Cambridge's best institutions and the draft Merrydown is notorious for its kick.

Another Cambridgeshire highlight was my visit on 13 August to Ely Cathedral, with Mary Thatcher, Pat Harwood, Peter and Adelma Longton (he was a lecturer in statistics with Jack Trevor), and Robin Savory from Pietermaritzburg. In my journal I wrote:

> The great beauty, gigantic dimensions and architectural brilliance of the edifice left me marvelling at the achievements of our ancestors. The infinite patience they were prepared to lavish on such a building – the Lady Chapel alone was a work of twenty-eight years! – contrasts sharply with the modern desire for quick results, high-pressure activities and minimum of effort [and I wrote this close on 50 years ago]. Has the machine age lost to us a sense of patience, of time-abiding tolerance? For these are the very virtues lacking in our science, our art, our politics. Results must be achieved quickly, in this lifetime: few have the happy phlegm of eternity and infinity in their outlook, that blessed state which I glimpsed and tasted in the timelessness and spacelessness of the Kalahari desert.
>
> 'What is this life if, full of care,
> We have no time to stand and stare?'
>
> The words of the poet William Henry Davies are so true, so desperately true, yet almost never in my young life – until this year – have I had the

> opportunity to stand and stare. This is something I knew the taste of – from my three-months' ramble in the desert – but not until my Cambridge visit has it been possible for me to take stock, not just of my work, but of life in general, to pour out some of these feelings on to paper, to look around me, appreciate the passage of the seasons, the changing climate of people and things; to be at ease and relaxed yet not idle.

My happy memories from 1955 are of picnics and serious discussions with such good friends as Noel Garson and John Reeves on Trinity Backs ... the angelic voices of the King's College Chapel Choir ... acres of blooming tulips in Lincolnshire in May ... Madrigals on the Cam in June ... the 'May Bumps Races' (held in June), which I attended with Jim Stewart and John Reeves ... the midnight matinée of *Between the Lines* by the Footlights Revue ... a stroll from Grantchester vicarage to Byron's Pool with Gertie and Brian Hudson ... supper with Bernard and Bridget Campbell ... dinners at High Table in Emmanuel College ... chatting with the veteran archaeologist of Cambridge, Miles Burkitt (not for him the bother of obtaining a Ph.D., but so brilliant a teacher and mentor that he did not need one to equip him to train many of the important figures in African archaeology) ... attending an open air concert on Trinity Backs ... and many other such delightful happenings.

Cambridge is indeed a marvellous place. But a year filled with only the social and cultural would have fulfilled only part of Lord Nuffield's purpose in setting up the Foundation Fellowships, nor would it have warranted inclusion in my autobiography. Let me turn now to some of my research pursuits during that year, one or two of which played a direct or indirect part in my future.

12 The Duckworth Laboratory

Jack Trevor, the director of the Duckworth Laboratory, was one of the world's leading experts on the methods of anthropometry (the measurement of humankind) and a scion of the Biometric School in London. He was also a gentle soul, a fey personality, and a paragon of erudition, with a ribald turn of mind and an astonishingly versatile collection of risqué limericks. Trevor was married to a charming Polish lady, Tusia, and I had some delightful dinner parties with them during my time in Cambridge.

During the Second World War, Trevor, who was already a physical anthropologist at Cambridge, joined the Intelligence Service of the British Army and was seconded to Kenya. There he became a good friend of the Leakeys and Louis Leakey spoke of him as 'my opposite number in military intelligence'.[7]

During his time as head of the Duckworth Laboratory, Trevor made it a centre for visitors from all over the world. There I met Professor Jon Steffenson, head of the Department of Anatomy at Reykjavik, Iceland; a group from the Soviet Union; Dr Lawrence Oschinsky, author of a book on the anthropometry of the Baganda and other Central African peoples and who was working in Burma (now Myanmar).

Trevor made a special study of race and the races of living humanity. He prepared an immense analysis of these subjects for an encyclopaedia in Israel. In 1951, the United Nations Educational, Scientific, and Cultural Organisation (Unesco) resolved to prepare a *Second Unesco Statement on Race*. Trevor was one of the seven physical anthropologists, along with Solly Zuckerman, J.B.S. Haldane and Geoffrey Mourant ('bony Mourant'), from England, and five geneticists, comprising Unesco's Second Committee. It met in Paris in June 1951.

The sinister shadow-show of race purity and the supposed evils of race-crossing were widespread not only in Germany but also in apartheid South Africa. Trevor was one of the few people to test the supposition that 'race crossing' was accompanied by 'harmful and excessive variability'. His study on *Race Crossing in Man: The Analysis of Metrical Characters* was published in 1953. His results did not corroborate the myths about the supposedly harmful effects of 'racial crossing'.

In the period immediately following the end of the Second World War, Trevor and Miriam Tildesley, another English anthropometrist, were drafted to the body charged with 'denazification' in occupied Germany. The tasks set for them by military intelligence were to interrogate each of the German physical anthropologists, and to determine what he or she had done during the period of 'race cleansing' in the National Socialist era of the Hitler regime. Many of Hitler's ideas on race and racism had indeed flowed from the writings of some German physical anthropologists. The works of men such as H.F.K. Günther, and his book, *The Racial Elements of European History*, provided grist to Hitler's mill.

Trevor and Tildesley had to clarify what part German anthropologists had played in the negative eugenical and genocidal programmes that had been developed to apply such ideas, especially to the mentally defective, the Jews, Sinti and Roma ('Gypsies'), the Christian Poles and homosexuals.

As the post-doctoral guest of Trevor in the Duckworth Laboratory, I heard many of the details of these experiences from him. Miriam Tildesley added to the first-hand accounts during her visits to Cambridge. I recall her recounting how, when the former anthropologist Von Eikstedt came for his interview, he was wearing his gauleiter's livery. The game and unremitting Englishwoman told him to go away and take off 'his Boy Scout uniform' before she would see him!

I soon became aware of the abiological stance of Trevor and so, while he imparted to me the elegant mental habits of biometrical and statistical analysis, I think I taught him something of the elements of genetical, developmental and evolutionary biology. We both enjoyed the interplay. Between us we achieved, I like to think, a limited interpersonal synthesis.

At that time I remained rather resistant to the biometrical approach. I wrote in my journal on 23 May 1955:

> ... the Biometrical School's approach was very liable to lead to the neglect of observation ... In their hands, a skull became simply an object

> of measurement and they were inclining not to look at the skull with their eyes . . . As Prof. Dart has put it, there is a danger of their becoming hypnotised by measurements. Not only that, but I find that there is a tendency for them to confuse taking measurements with doing research or applying scientific method. One of them thought he was doing research when he was making measurements, until I asked him whether research is being carried out by a tailor who measures a client! Or a hatter fitting a hat! It is clear that scientific method is apt to be lost sight of, in biometric procedure, possibly because of a tremendous preoccupation with methodology, both metrical and statistical . . .
>
> I used to believe there were two types of scientist: the cautious type whose hypotheses are ahead of the facts by not more than a fraction of a centimetre; and the bold, courageous type whose hypotheses stand out like signposts miles along the highway, pointing a way in the darkness. (Lawrie Wells and Berry Malan are examples of the former; Raymond Dart and Louis Leakey of the latter.) It now seems I must admit a third category, he who makes no hypotheses at all! – since measurement and statistical analysis have become a substitute for thinking.

Today, my ideas have mellowed somewhat and moved a little away from the pattern of opposites that struck me 50 years ago, although the computer seems now to have taken over from the brain, from thinking and from developing hypotheses, for many of my colleagues and students.

I return to Jack Trevor. Shortly after my 1955 visit, his marriage broke up and he went into a depressive decline. In 1964, when he was often unable to fulfil his lecturing duties, I was invited by the Faculty of Archaeology and Ethnology to come to Cambridge as a visiting professor. My task was to take over his lectures and supervisory sessions with students.

My intellectual adventures on that 1964 visit revolved around *Homo habilis* and I shall come back to that later (see Chapter 31). When Trevor came to the Duckworth Laboratory to see the Tanzanian fossils, little did I think it was the last time I would see him. Three years later, David Pilbeam went to visit him. Unopened milk bottles had accumulated outside the front door. Pilbeam clambered through the lounge window and found Jack's lifeless body in his bedroom, half in and half out of his bed.

I look back with affection and gratitude to Jack Trevor who introduced me to Cambridge and opened my eyes to another facet of physical anthropology.

I HAD BARELY been installed in the Duckworth Laboratory, when Trevor said, 'Get your coat and hat; we're going to Sir Arthur Keith's funeral'. Keith had died a few days before and his ashes were to be buried in the grounds of Down House in the village of Downe. I had never met Sir Arthur Keith though I knew a good deal about his role as the doyen of anatomists and anthropologists in the British Isles for much of the first half of the twentieth century.

I fetched my coat and Homburg and off we went, taking two trains and a taxi to arrive in time. It was 3.00 p.m.; the sun stood only about fifteen degrees above the horizon, glowing pale orange in the subfusc atmosphere. It was bitterly cold. About a dozen men in black coats and hats stood around, their breath visible, stamping their feet against the chill. Jack Trevor stood next to me and whisperingly told me who some of the mourners were. The memory, like the misty air, is nebulous, but I imagine that Sir Wilfrid LeGros Clark, Dr Joseph Weiner, Professor A.J.E. Cave, Dr Kenneth Oakley, Sir Gavin de Beer, Sir Solly Zuckerman, as well as dignitaries from the Royal Society, the Royal College of Surgeons, and the Natural History Museum, must all have been there. The little casket was laid in the earth; there was no service (Keith was a free thinker). The party broke up and we went back to Cambridge.

I mention this occasion as I have not traced any published account of that little ceremony and on a recent visit to Down House found that even the spot where the ashes were inhumed had not been marked.

ONE OF THE projects I undertook in 1955 on Jack Trevor's suggestion was a study of the curvature of the occipital bone. This is the bone that forms the back of the cranium and lies over the occipital lobe of the cerebrum and the cerebellum. I approached this intriguing area from the points of view of methods to determine the curvature, variation in adult modern human beings of many different population groups, changes in the curvature from babyhood to adulthood, and a search for changes in the curvature during the course of hominid evolution. It was a useful exercise in craniometric analysis coupled with the biological underpinning, and six published articles stemmed from this project.

In June 1955, the mail brought the good news that the Council for Scientific and Industrial Research (CSIR), in Pretoria, had accepted the final agreement for the publication by Messrs. Percy Lund, Humphries of London, of the book based on my Ph.D. thesis. My first book was to be entitled *Chromosomes, Sex-Cells and Evolution in a Mammal.* After a number of years of delay, work on the typesetting and printing could begin. The first proofs arrived before I left Cambridge at the end of 1955. As I spent the next six months in the USA, batches of proofs followed me by airmail, and often had to be corrected late at night.

The foreword was kindly written by Dobzhansky, then at Columbia University, New York City. His words were encouraging:

> Many patient and critical observers have studied the structure and formation of the sex cells in most diverse organisms. Dr Phillip V. Tobias is one of these observers who has made a fine contribution to the biological knowledge by his painstaking and detailed work on the male reproductive gland of the South African gerbil. The results of these important studies may be summed up very simply: unity within diversity.

I especially appreciated the next sentence: 'An extreme simplicity is characteristic of most of the really important conclusions in science!' My first book appeared in 1956 with a small print-run of 500. I was extremely gratified when it was quickly sold out.

I HAD MADE up my mind that, if I were going to become a palaeoanthropologist, I should try to examine every fossil hominid available. The purpose was simply to familiarise myself with the morphology and variations of ancient hominids.

The Natural History Museum in London is one of the half-dozen greatest repositories of fossil hominid remains in the world. At the kind invitation of Kenneth Oakley and with the help of his assistants Theya Mollison and Rosemary Powers, I made fairly frequent visits from

Cambridge to London, to familiarise myself with the original hominid fossils, especially those from Africa.

One of the most magnificent crania was that from Kabwe or Broken Hill (formerly known as Rhodesian Man) on the Copper Belt of northern Zambia (formerly Northern Rhodesia). This very complete cranium with diseased teeth and gums was the first significant hominid discovery from Africa. It was obtained in 1921, three years before the Taung skull was found in South Africa.

During May to June 1955, I also examined some fossil hominid specimens that Louis Leakey had recovered in western Kenya in the early 1930s. The sites of Kanam and Kanjera lie on the north shore of the Gulf of Kavirondo that protrudes, finger-like, from the north-eastern corner of Lake Victoria. The specimens and the positions of their discovery had been the cause of a major to-do that, before the Second World War, had threatened to torpedo Louis Leakey's reputation. I knew a little about this history at the time and so it was with particular interest that I examined the Kanam jaw fragment and the Kanjera cranial pieces. One of the features of the jaw that had, in the minds of some investigators, disqualified it from being of any high antiquity, was its possession of a supposed well-developed chin.

It is a fascinating problem in hominid evolution that all ancient hominid specimens lack what we call a bony chin. That is to say, the front of the jawbone does not project forwards; instead, the front of the mandible in the midline retreats below the front teeth (the incisors and canines). By contrast, recent and modern humans, of the kind we call *Homo sapiens sapiens*, have a definite bony chin, protruding anteriorly to a greater or a lesser degree. Kanam was an enigma: Leakey was claiming that it was of a great age, whilst it supposedly showed the modern feature of a substantial chin. This presumed 'chinniness' was one of the reasons, though not the only one, why the world was suspicious about the claimed great age of the jaw.

Almost from the first moment I examined the Kanam jaw, it struck me that the so-called chin was off-centre and lay over a pathological swelling of the bone. Could the 'chin' be nothing more than an overgrowth of bone across the front of the pathology? The diseased state of the bone had been interpreted by the pathologist of the Royal College of Surgeons, T.W.P. Lawrence, as a 'sub-periosteal ossifying sarcoma', a kind of bone tumour.

Duckworth had considered the supposed chin to be an exostosis. To prove the point, I needed to make a complete re-study of the jaw fragment.

What I saw when I examined microscopically a section through the swelling was enough to convince me that it was not a bone tumour or sarcoma. A fracture was evident and it was possible that the swelling resulted from an infection (osteomyelitis) of the jaw with a massive bony reaction to the chronic infection. Some compensatory bone growth on the front surface of the jaw simulated a chin.

I published a summary of my results in *Nature* (in 1960). In 1959, at the Pan-African Congress of Prehistory and Palaeontology, at Kinshasa, I presented a detailed account. While I disproved the evidence on which Leakey had erected a new species *Homo kanamensis*, my analysis largely cleared Leakey of the charges and aspersions which in the late 1930s had threatened to blight his international scientific reputation.

At that 1959 Pan-African Congress in Kinshasa, Louis Leakey presented the first account of the important new cranium of Olduvai hominid 5, which Mary Leakey had discovered only five weeks earlier. At the same meeting Louis and Mary took me to one side and invited me to undertake the description of this remarkable new cranium. That story and how in a trice it changed my life's direction is for telling another time.

13 French interludes

As part of my plan to examine at first hand as many original fossil hominid specimens as I could, I had worked my way systematically through the collections in the Natural History Museum in London. In April 1955 I was fortunate to make a first familiarising visit to France. In the Musée de l'Homme in Paris, I was received with great kindness by the director, Henri V. Vallois, and the sub-director, Peter Champion, as well as Monique de l'Estrange, who had previously translated my articles on the Kalahari San (Bushmen) into French for publication in *L'Anthropologie* (see Chapter 10).

I was given a private office and laboratory. Monique provided me with instruments and the skulls I wished to examine. Over the ensuing days, with an interlude in the south of France, I was able to study the most important fossil hominid holdings in that great museum: some that were classified as Neandertal (La Quina, La Ferrassie, La-Chapelle-aux-Saints and Malarnaud) and some as 'Cro-Magnon types' (Cro-Magnon, Duruthy, La Madeleine, Arcy-sur-Cure). In addition, I studied fourteen adult Khoe-San skulls in the Musée de l'Homme.

Vallois took me across to the Muséum National d'Histoire Naturelle and introduced me to Professor Camille Arambourg. Doyen of French palaeontologists, Arambourg was a neat, dapper figure. He spoke French to me, whilst I, not as fluent as I should have been, spoke what a humorist called 'fractured French'. It was an interesting thing about Arambourg: when he saw me struggling for the right word in French, he made no attempt to help me out – although he knew what I was trying to say. He sat patiently waiting and watching as I strove to express myself.

My visit to Arambourg was memorable for two reasons. While I was chatting with him, he received a package from Morocco. He unpacked it before me and I was thrilled to see that it contained an interesting hominid jawbone from Sidi Abderrahman. It was Arambourg's – and my – first sight of this newly discovered fossil.

The second reason why I remember that visit to Arambourg was that he took me down to one of the exhibit halls and there he showed me, among other things, some of Sarah ('Saartje') Baartman's remains on display. Apart from her well-preserved, articulated and mounted skeleton, as far as can be ascertained only two parts of her soft tissues had been preserved: her brain and her external genitalia. They were kept in embalming fluid in sealed glass jars. There they were on public display in the Muséum National in 1955. I remember that these 'soft tissues', as anatomists call them, were discoloured and darkened, and detail was scarcely detectable.

I did not know at that time, 50 years ago, the sad and sordid history of the life and death of the Khoe-San woman Sarah Baartman: my knowledge of that was to come later. The reason why Arambourg wanted me to see these remains was that he was aware that I was deeply involved in studying the living San peoples of the Kalahari desert and had been a member of the French Panhard-Capricorn expedition (see Chapter 10). Perhaps also he saw the remains of this Khoe-San lady as a link with my country.

I gazed at these publicly exhibited remains and myriad reflections welled up within me. Most of the skeletal remains labelled 'Khoe-San' in museums and university departments had been excavated from archaeological deposits, and they were anonymous. We could say that the cultural remains found with them were of the Wilton or Smithfield or other industry of the Later Stone Age, and we could act as 'medical detectives' identifying the age, gender and physical features of the deceased. But rarely could we say, these bones belonged, for example, to Mrs Waterboer or Mr Khomani. In that sense they have remained mute and nameless. By contrast, the remains that the Muséum National were exhibiting bore a name. This was a known-in-life individual, identifiable not only by age, gender and ethnicity, but by the name she bore and by the date of her death. Who was there today to mourn over these remains and to pay homage to her memory?

The full horror and personal tragedy of Sarah Baartman was later borne in on me when I studied all that I could lay my hands on about her life in South Africa and Europe. The enticement of Sarah Baartman to England, to exhibit her body and supposedly become a rich woman, exemplifies the biases of the day towards subject peoples. It incorporates attitudes of racial superiority and inferiority, and of a prurient sexism, for the London and Paris audiences who paid to see her were inordinately curious about her sexual features – as indeed were the French scientists, especially Georges Cuvier. The history of this Khoe-San woman epitomises an entire epoch of racial repression and colonial subjugation.

From 1996 to 2001, at the request of the South African government, I carried out protracted, often tedious and frustrating negotiations for the repatriation of Sarah Baartman's remains, especially with the museum director, Dr Henry de Lumley.

In petitioning the French authorities and the Muséum National d'Histoire Naturelle for the return of Sarah Baartman's remains – or such of them as were still available – we (the South African authorities and academics) assured our French colleagues that we had no intention to embark upon a campaign for the repatriation of other cultural and skeletal remains from the scores of museums, galleries and universities in most European countries and other places where such objects and remains repose. We were well aware that there are literally thousands of skulls, skeletons and cultural objects, which were removed from South Africa during the colonial era, almost always without informed consent being given by the indigenous peoples closely concerned. These thousands of objects are to be found in many of the great museums of Europe, North America and Australasia.

However, we affirmed that we believed that the case of Sarah Baartman was special and, indeed, unique. The remains of Sarah Baartman were of an individual whose identity was known in life, who was baptised, married and the mother of children, and about whose life-history many details are known.

Finally, in a three-stage procedure, the French parliament in January and February 2002 approved a Bill to permit the return of Sarah Baartman's remains as requested by the South African government. The remains were handed over by representatives of the French government to the South African Embassy in Paris on 29 April 2002.

I spoke at the reception of her remains at Cape Town International Airport on 3 May 2002. Three months later, Sarah Baartman's mortal remains were received by another large gathering at Port Elizabeth Airport in the Eastern Cape province.

The following day, 9 August 2002, coinciding happily with South Africa's National Women's Day and the International Day of the World's Indigenous People, her remains were laid to rest at Hankey, a little distance inland from Port Elizabeth. The burial was attended by President and Mrs Thabo Mbeki, Minister Dr Ben Ngubane and thousands of others. I could not help reflecting how attractive the site is, on a hillside overlooking the village of Hankey, near where Sarah Baartman had been born about 213 years earlier. I still hope that my proposals for the erection of a suitable tomb may come to pass.

IN APRIL 1955 I visited a third Parisian centre, the Institut de Paléontologie Humaine, of which Henri Vallois was also director. On this and subsequent visits, I was able to examine the tantalising Fontéchevade cranial fragments, that of Djebel Kafzeh from close to Nazareth, Galilee, in Israel, and from Africa the mandibular fragments of Dirè-Dawa in Ethiopia and Rabat in Morocco, the skull of Asselar in Mali, and twenty of the Afalou-bou-Rhummel skull fragments from Algeria. I also saw a San cranium said to have been presented by the Revd John Campbell. He had ministered to the Griqua village of Campbell early in the nineteenth century and a small group of Bushmen were said to have been living in a cave in the gorge close by. Could this have been the source of that lonely skull in the Muséum? How many hundreds of Khoe-San crania and skeletons were purloined from southern Africa and now repose lonesomely in institutions all over the world?

Perhaps the highlight of my 1955 French visit was a three-day trip to the Vezère Valley with Henri and Mme Vallois from Paris and the archaeologist Malvasin-Farbre of Bordeaux. The cave sites I visited are too numerous to detail here, but they included the type-site of Cro-Magnon, and Le Moustier from which the juvenile Neandertal skeleton was recovered by the Swiss antiquary Otto Hauser in 1908 (see Chapter 32). I visited also La Ferrassie from which a number of early human remains had come to light, but, Vallois told me, had not yet been published.

The final cave visit was to Lascaux, that superb subterranean art gallery that had been discovered in September 1940 by five schoolboys of Montignac out walking with their dog. It disappeared. Their cries and whistles elicited a faint underground barking. Finding an opening like that of a rabbit warren, the boys widened it and slid in one by one. As they lit matches they found themselves in a great cavern, from the walls of which large beasts leapt at them in the flickering light. Thus was Lascaux discovered – with its trotting ponies, its falling horse, its swimming stags, its charging bison, its leaping cow and its five-metre bulls.

It was just at that time, in 1955, with about 1 000 visitors a day to the Lascaux cave, that a serious threat to the survival of the paintings was detected. As Vallois pointed out to me, fungus had gained access to the interface between the paint and the natural, translucent, calcite glaze that had formed over thousands of years. As the fungus spread deep to the glaze, the paint was being 'consumed'. The body heat and humidity of people pressure had so changed the environmental conditions within the cave that they had allowed free rein to the fungus. Even the laying out of bowls of hygroscopic substances had failed to restore the intracavernous atmosphere, nor was the exhibition of antibiotics deep to the glaze wholly successful in stemming the advance of the fungi. There was nothing for it but to close the cave to the public and our visit was taking place in the last week that the cave was open. I noted in my journal on my visit to Lascaux, 'Have never been so deeply moved by such a site before.'

Fortunately, the paintings have been saved. To cater for the public's interest, a replica of the cave and its contents has been created close by and it is apparently as popular with visitors as the original had been.

MY SECOND TRIP to Paris in 1955 was to represent the Wits Anatomy Department at the International Anatomical Congress held from late July to early August.

The Congress provided my first opportunity to participate in a meeting of the International Anatomical Nomenclature Committee. My linkage with this body and its successor, the Federative International Committee on Anatomical Terminology, lasted from that date (1955) until I stepped down in 2004. It brought me into close contact and friendship with a dozen or more leading anatomists from all six continents. We have met on average twice a year.

It is a universally accepted truth that the most important things that happen at international meetings and congresses are not so much what you learn in the sessions and lectures, but rather the friction of mind upon mind outside the sessions. It is there that you make new friends and renew old acquaintances. There joint research undertakings may be planned, reciprocal visits and lectures arranged, and beyond all of these serious-minded doings, are the dining and wining among like-minded people. Paris, 1955, was no exception. Among others, I met many genial American professors, a number of whom invited me to visit their departments in 1956 – for already I was thinking of a sponsored six-months' visit to the USA after the end of my stint at Cambridge. I had begun negotiations with the Rockefeller Foundation, whose Paris offices I visited during the Congress.

A quite extraordinary episode occurred during the meetings of the physical anthropological section of the Paris Congress. This sectional meeting was very well attended and my little paper was satisfactorily received. Then it was the turn of Professor (later Sir and then Lord) Solly Zuckerman, who was head of the Department of Anatomy at Birmingham, England. For years Zuckerman had been conducting a veritable vendetta against Dart, Broom and *Australopithecus*, and their claims that it showed some hominid features. In Paris, I had occasion to disagree publicly with the conclusions he expressed in his paper. The fat was truly in the fire.

A young 29-year-old upstart had dared to challenge the very senior and authoritative Professor Zuckerman. A number of people applauded my intervention; others seemed to be too nervous or too embarrassed to do so. Somebody from Glasgow whispered to me, 'Now you've lost any chance of getting a chair in Britain.' This was an allusion to the immense power that Zuckerman was already wielding in British academic circles, including his service as scientific adviser to Whitehall and his impending knighthood. But I was certain that he had drawn the wrong conclusions from his data.

Zuckerman jumped up and it was clear that he was absolutely livid. He delivered himself of a diatribe, which was more personal than scientific. Having trampled me underfoot, he sat down and sent me a hastily scribbled note: 'Three cheers, Tobias. You did well to keep the Dart flags flying!' He implied thereby that my strong criticism had been made purely out of a sense of loyalty to my chief. I sent a little note back to this effect: 'Thank you, Professor Zuckerman, but I believe that what I said was correct.' At the end of the session, as people were trooping out, Zuckerman buttonholed me, gave me a slap on the back, and invited me to have a drink with him!

I HAD FIRST visited Zuckerman at Birmingham (at Dart's instigation) at the time of the great fog in London early in 1951. I invited Zuckerman to visit the Anatomy Department at Wits, saying that we would give him unfettered access to the fossil hominids. I added, 'You'll be overwhelmed by the evidence.' Immediately he replied, 'That's just it: I don't *want* to be overwhelmed!'

Although he had declined to come out to see the Taung skull and the other australopithecines, when he needed some tooth measurements of the Taung child, Zuckerman did not hesitate to ask Dart, who bade me give him what he wanted. With a great sense of occasion – for I had never before seen or handled the original child skull – I carefully measured the teeth requested by Zuckerman. The measurements went across and it was these that he used in a paper in *Nature* in 1950 purporting to show that the canine teeth of Taung were no different from those he had measured in juvenile apes. This paper featured a serious error: in working out his statistics, he omitted to divide by the square root of two! Zuckerman subsequently acknowledged his error, but went on to claim that correcting it made no difference to the results.

It was only in 1975 that Zuckerman at last visited South Africa, primarily to give the Cecil Rhodes Lecture at Rhodes University in Grahamstown. He accepted an invitation to give a talk at Wits and I wondered whether he would use the occasion to attack Dart's work on *Australopithecus*, in the very home of *Australopithecus*. However, he gave an excellent address on the organisation of science in the UK, the USA and the USSR. From his years of involvement in Operation Research and as a senior scientific adviser to the UK government, nobody was better placed to make such a comparison.

Over a glass of sherry after the lecture, I re-introduced myself to him and to Lady Zuckerman and renewed my invitation to visit the Anatomy Department and see some of the fossil hominids. (We had by then obtained hundreds of new specimens from Sterkfontein, to supplement the Taung skull and a few dozen specimens from the Makapansgat Limeworks.) He replied that he would have to see if he could fit such a visit into his schedule. A day or two later, he phoned my secretary and said, 'Tell Professor Tobias I can give him an hour on Thursday morning.' He spent not one but three hours in the Anatomy Department. This gave my assistants and me time to show him a good selection of our hominids. Alas, it seemed that his mind was closed. Every few minutes he would point out on one of our

Australopithecus fossils what he declared to be a feature of chimpanzee or orang-utan, or even gorilla. He remained unconvinced that these australopithecine creatures were hominids or close to the hominids, and his visit to the Anatomy Department did not subsequently rate even a mention in his autobiography.

BEFORE THE 1955 Paris meeting was finished, we enjoyed a fine reception in the Hôtel de Ville. Champagne and refreshments were served in the sumptuous halls, gilded, ornamented, mirrored and painted in the most extravagant French Baroque style. We could study surface anatomy in the Hall of Sculptures. Suddenly, after three and a half days in Paris, I found myself tired and alone amid all the wonders and the friendliness. I wandered out into a rainy night, wheezing gently, across the courtyard, past the Petite Arc de Triomphe, and down the long, broad walk of the Jardin des Tuileries, ill-lit, and flanked by thick, lightless woods. Among the trees, solitary shadowy figures flitted, keeping – I imagined – a night's vigil on the undercover world of the dark, menacing yet intriguing, and keeping their distance from the flirting couples who were braving the wet. And so I wandered out on to the mighty Place de la Concorde, its fountains frolicking in powerful arc lights, its obelisk floodlit so as to give it a squat, square-topped aspect, the surmounting pyramid being lost in the shadow from below, as though a great culture had lost its head in the dark, truncated abruptly even as it aspired to proud heights.

The romance of the city entered into my soul and I was no longer lonely. What good company the spirits can be! How they can crowd in upon you and fill your consciousness with their rich vitality. How vibrant the spirits of Paris are, in truth.

Then as the bronchospasm tightened on my chest, I passed by Metro from Concorde along the Vincennes-Neuilly line, 'Direction Neuilly' as far as the spacious square of Porte Maillot, separated by the traffic-crowded Avenue de la Grande Armée, from Etoile and the Arc de Triomphe; and the 82 bus from there to Perronet. The smell, the redolence, is rich and strong and strange in the subways, on the crowded Metro coaches; and French eyes are quick and observant and searching, always searching. The sweltering heat cooked up the smells and the eyes and the people and the movement and the clanking into one sensuous blur – Paris. Essence de Paris: lottery stalls and pissoires on the pavement, wine on the sidewalk, news-stands where, without buying, queues of people stand and read the latest news headlines, questing eyes meet and people kiss on the sidewalk.

And so, back to Cambridge.

14 Last months in Cambridge

I had been working in the Duckworth Laboratory for six or seven months, when one day I received a little note, written in courtly style, inviting me to take sherry with Dr Wynfrid Laurence Henry Duckworth in his rooms at Jesus College, further along Jesus Lane than my lodgings. Our first rendezvous on 8 August 1955 marked the start of a series of visits that continued until the end of my sojourn at Cambridge. I last called on him when he was a patient in the Evelyn Nursing Home on 26 December 1955.

Duckworth had been a member of the Cambridge Anatomy Department from 1898 until 1940. It may well be asked then how it came about that his name was given to a laboratory, not in the Anatomy Department, but in the Faculty of Archaeology and Anthropology. Apparently two factors were involved. In his autobiography, Arthur Keith relates that when Alexander MacAlister, the professor of anatomy at Cambridge, died in 1916, he (Keith) was one of the electors to choose a successor. Duckworth was seen by many as an automatic successor to his old mentor, MacAlister. However, there was a feeling by some, especially the Cambridge physiologists, that what was needed was not a person who was primarily a physical anthropologist (as was Duckworth), but one conversant with modern anatomical research methods. Elliot Smith of Australia was then working at Manchester and was to become the director of the new Institute of Anatomy at University College London, a few years later. He favoured the view that Cambridge needed a modern anatomist, and he persuaded his and Dart's old master, J.T. Wilson of Sydney, to apply. So, notwithstanding his long friendship with Duckworth, Keith voted for Wilson. Duckworth suffered a near-mortal blow. He had set his heart on the Cambridge chair and his hopes were dashed. As a consolation, he was made a reader in the Cambridge department and became master of Jesus College.

Throughout his lengthy and wide-ranging conversations with me, he did not once raise this subject. But I was told that, once a year, during a lecture to the anatomy students, he would look out of the window and say something of this nature: 'And so gentlemen, if on just such a fine spring morning as this, you awaken to find that your life's highest hopes have been irreparably frustrated, do not lose hope, gentlemen, even though it may be difficult.'

There was another problem involving Duckworth and the Cambridge Anatomy Department. The head of the department in 1955, J. Dixon Boyd, told me that Duckworth had largely followed the old 'dry-as-dust' kind of anthropology, and this had become something of an incubus in the department. At the other end of the Downing Street site, in A.C. Haddon, there had been a general anthropologist who was very interested in physical anthropology. So physical anthropology gravitated over to the Faculty of Archaeology and Anthropology and took the name of Duckworth with it to the laboratory.

Incidentally, the Oxford Anatomy Department had likewise reacted against the 'old, dull and unprofitable' anthropology and had also thrown it off, sending its skulls down to the Natural History Museum in London. Boyd said, 'It was Woollard and LeGros Clark who led the emancipation of anatomy.'

I found Duckworth in his rooms at Jesus College, 85 years old, living in comparative retirement except for occasional tutorials. He told me how he had devoted a large part of his life to Jesus College and that the College had been very good to him, giving him a suite of rooms and other privileges for the rest of his life.

His memory, he told me, had gone quite suddenly about six months ago, since which time it was very bad. It seemed to come upon him during the conversation, almost in spells or flashes. Every now and again, something would swim into his memory and he would give utterance to it – hoping, in the most charming way imaginable, that he was not being too incoherent or tedious. He told me much that was stimulating and fascinating and I jotted down a few notes.

On my second visit to Duckworth, I found that he had rallied and moved to his study. There, on the desk before him, were numerous reprints and boxes of papers. He gave me a number of these to keep, telling me in each case about the origin of the material and what he had made of it. All the time, in the most gentle and humble manner, he would ask whether 'I would care to . . .' or '. . . be good enough to' take such-and-such a reprint, and whether I would be good enough to quote this or that, where relevant to my own work.

He seemed to have developed a new accession of liveliness as we chatted, listening avidly as I told him about my views on the Khoe-San and on African evolutionary problems, and warming to his theme as he explained the various reprints he gave me. I could readily see and believe how a retention of interest and a maintenance of faculties go hand in hand, since his memory seemed to be improving progressively – as also his speech – as the conversation waxed over a sweet sherry.

We met on a few other occasions and I always enjoyed our conversations. Late in December 1955, I wrote to Duckworth requesting a farewell interview. After a lengthy delay, his maiden sister replied on his behalf saying that he was too ill to write and had been taken to a nursing home. A few days later, his sister arrived with the request that I visit him in the nursing home. On 26 December I found him considerably deteriorated in body and mind, but still with a flicker of spirit that saw him detail books and papers that he wanted me to look through and sort for him.

I sailed for New York on 12 January 1956. On my arrival there I received a telegram with the sad news that Wynfrid Duckworth had passed away on 15 January.

TWO DAYS AFTER my birthday in October 1955, I drove across to Oxford at the invitation of Sir Wilfrid Edward LeGros Clark, the professor of anatomy known widely and affectionately as 'LeGros' (the final 's' is not pronounced as in most French words ending with 's').

LeGros was friendly and helpful and offered me the facilities of his department during the week of my stay. We exchanged ideas about the South African fossil hominid sites and the specimens that had so far emerged from them.

LeGros and I also spoke about my future. He advised me to specialise. The time had come to take a definite decision between cytogenetics and physical anthropology. I told him that my book *Chromosomes, Sex-Cells and*

Evolution in a Mammal, was due to appear the following year. Apart from that I had done very little cytogenetics for some five years. In effect, I had already specialised in anthropology. He was pleased and said I should go deeper and deeper into this and become an authority on one branch. I told him I was already specialising in the study of the living peoples of Africa, especially what I proposed to call The Old Yellow South Africans or the Khoe-San. I had studiously avoided doing anything on the australopithecines, as I felt that this was Professor Dart's special preserve, at least as long as he remained active in the field. LeGros agreed that this was the correct thing to do and wished me all the best in my future career.

One milestone of that career occurred on 25 November 1955, when I gave my first paper at a meeting of the Anatomical Society held in the Department of Anatomy of the Middlesex Hospital Medical School. The meeting was running late after lunch and my presentation had to be postponed until after tea. My paper was on 'Premeiotic mitosis: A new type of cell division in mammals'. I had some projection slides and spoke without notes, but there was no time for questions and discussion after the paper. Instead, they were deferred until the end, but I had to leave before that to catch my train back to Cambridge. I was disappointed. Until you hear a little discussion, you do not really know whether you have put a topic over or not and my impression was that I hadn't spoken as well as I should have liked to.

By chance a night or two later, I was reading Arthur Keith's autobiography and came across an account of his own first paper to the Anatomical Society in the spring of 1893. He states, 'My diary records that "I made a terrible mess of it."'

I had just turned 30; Keith had been 27. Then I remembered the wretched time that Raymond Dart had had when he demonstrated the original Taung skull to the Zoological Society of London, on 17 February 1931. Apparently Elliot Smith had presented an account, illustrated with lantern slides, on Davidson Black's new discovery of a superb Peking Man skull. His was 'a masterly demonstration' (in Dart's words), but Dart called his own presentation an anticlimax, 'a fumbling account' and 'a pathetic unrehearsed showing'.[8] He had just turned 38 a fortnight before.

Although I was personally disappointed at my presentation to the Anatomical Society, there was one gratifying consequence. As I was leaving to run for my train back to Cambridge, Professor George Romanes, head of the Department of Anatomy at the University of Edinburgh, detached himself from the audience and slipped out to have a word with me. He offered me a position in his department as a senior lecturer and named an attractive salary. He wanted a physical anthropologist, liked my approach, and was sick and tired of skull measurements (with which I heartily concurred). He did not want to rush me into a quick decision, but would leave the thought with me. I took my train back to Cambridge in a state of high excitement, the disappointment of having no discussion on my paper mingling with the prospects raised by Romanes.

DURING MY YEAR at Cambridge, the students there were becoming more conscious of the evils of the apartheid policy in South Africa and of its consequences. Ambrose Reeves, the anti-apartheid Bishop of Johannesburg, was invited to speak to one of the societies. A large audience heard him vigorously attacking the South African government's policy. The South African Students' Association at Cambridge invited the information officer at South Africa House in London, Mr A. Steward, to give a talk on apartheid, presumably as a counterblast to Bishop Reeves's address. I was among the audience in the Cambridge Union and listened in dismay as Steward tried to justify the apartheid policy. The burden of his address was to try to show that the whole purpose of apartheid was the upliftment of the black people!

In the discussion from the floor, I could not resist rising and speaking. I showed that the purpose of apartheid was not at all as stated by the speaker. It was to enable the whites to retain power and to save the whites from the supposed threat of the blacks. To bolster my argument I had come armed with two or three quotations from speeches by Nationalist prime ministers and other ministers. I went on a little longer than is customary in the Cambridge Union, but the chairperson made no attempt to cut me short. I made it clear that I was speaking as a South African who had studied the growth of the apartheid policy since the elections that had brought the National Party to power seven years previously. My remarks were very well received by the audience and I noticed the speaker asking the chairperson (a South African) who I was.

Word of my intervention must have reached the seat of power in Pretoria. At least that is what I inferred when I was forced to endure insufferable delays after I attempted to have my South African passport endorsed to enable me to use it for my visit to the USA at the beginning of 1956. The Rockefeller Foundation had awarded me a Rockefeller Senior Travelling Fellowship to enable me to spend six months in the USA, from January to June 1956. It was to be my first visit to America.

Permission was delayed by Pretoria. I applied at South Africa House on 29 October 1955, and was told the process would take six weeks. I sat and waited for a nerve-racking month beyond the six weeks.

I was in what came to be called a 'catch-22' situation, since I could not apply for my US visa without the appropriate endorsement from Pretoria in my South African passport. Eventually I was forced to write to the principal of Wits, Professor W.G. Sutton, and to cable him. His reply telling me the endorsement of the passport had been approved came through four days before my ship sailed for America. I had to make another trip to South Africa House. By then I was reduced to a frazzle. On Monday, the day before sailing, I departed from 7 Jesus Lane, which had been my home in Cambridge for almost a year. I had packed, arranged for the dispatch of my books and papers to Johannesburg, sold my car at the last moment, picked up my passport endorsement at South Africa House and my American visa at the US Consulate in Grosvenor Square, and entrained for Southampton to embark on the *Queen Mary* the next day.

My interpretation of the way I had been treated by the South African government was that it added up to sheer vindictiveness and intimidation, for my anti-apartheid activities and no doubt for my unbridled remarks in the Cambridge Union. There is little doubt in my mind that the reason that I was reduced to a state of nervous prostration was to warn me and make a thinly veiled threat, something that others had suffered too. It was a couple of days out on the black, rolling North Atlantic, before my equilibrium started to be restored.

15 | Arriving in the USA

I went directly from England to America, thus extending my absence from South Africa. The *Queen Mary* was nearly three times the size of the Union-Castle liners with which I had plied my way between Cape Town and Southampton on previous occasions. In these days of jet-propelled intercontinental and interhemispheric travel, when people feel seriously put out if the flight to their destination exceeds ten hours, it is well to remember those leisurely journeys. Thirteen days were needed to get from South Africa to England or five days from Europe to North America. I once corrected a set of proofs and wrote a long article between Cape Town and Southampton!

On 17 January 1956, the *Queen Mary* docked at 7.00 a.m. in the pier at 50th Street, New York City. I was overwhelmed and dizzy to be received by the amazing skyline of Manhattan – the Chrysler Building and the Empire State and the Rockefeller Center were already there, but not yet the tragically short-lived twin towers of the World Trade Center. Snow was falling as we arrived, though it had let up by the time we were through the Customs formalities.

I was met not only by the skyscrapers and the pattering of snow, but by a representative of the Rockefeller Foundation, Theodore Fedewicz. He assisted me through the formalities – and right up to my bedroom in the Hotel Abbey on West 51st Street, half a block away from the Rockefeller Center. There, in the afternoon, I was received by Dr Gerard Pomerat who was in charge of my programme.

My visit to America was very different in nature from the Nuffield Fellowship at Cambridge. There I had been ensconced for a full year at one academic centre, with half-a-dozen side-trips around England, Scotland and France. My American award, by contrast, was a Travelling Fellowship, and in a six-month period I visited some 30 university departments, museums and research centres. A summary of all the people, places and departments I visited from the east to the west and back again, and from the Canadian to the Mexican borders would be tedious. In six months I filled three volumes of my ever-present journal (in contrast with one volume in twelve British months), and in these pages I relate only the highlights of my experience. The principal focus was on my three major academic interests: anatomy, physical anthropology and genetics, especially human genetics.

During my first four days in New York City, I saw high places: Manhattan bathed in sunshine from the top of the RCA building (70 storeys, then the third highest) and the fabulous spectacle of Manhattan by night from the top of the Empire State building (102 storeys). I also attended for a few minutes a meeting of the Security Council in the United Nations building.

Then, sufficiently primed and exhilarated, I fled by the Pennsylvania Railroad to the more genteel milieu of Baltimore. The journey took three hours with snow-covered countryside shimmering silently past.

A highlight of my visit to Baltimore was an afternoon spent in the Embryology Department of the Carnegie Institution. I was shown around by a gracious, white-haired, 70-year-old George W. Bartelmez, who had formerly been professor of anatomy at the University of Chicago. He allowed me to see one of the classical specimens in human embryology, the Hertig-Roc embryo of seven-and-a-half to eight days old. At one time this was the youngest human embryo.

I looked at the Hertig-Roc embryo under the microscope for a long time, trying to determine whether it was male or female. To do so, I applied a new method that I had devised based on the complement of nucleoli. With the kind co-operation of J. Dixon Boyd of the Anatomy Department at Cambridge, I had applied the method successfully to embryonic liver cells: Boyd's embryos were old enough for the gonads to have differentiated into either ovaries or testes. I studied their liver cell-nuclei 'blind', that is, without being shown the gonad or told the sex as Boyd determined it. Of a dozen embryos whose nucleoli I had examined at Cambridge, I correctly determined the sex in eleven. In the twelfth, the histological fixation was too poor to apply this new method. In the Carnegie collection, it seemed to me that the Hertig-Roc embryo was a male, as it had only three nucleoli per cell, although in some of the very large trophoblast cells, these nucleoli were greatly enlarged. The beautiful cytological detail in the Hertig-Roc embryo, and the hundreds of others in the collection, had never been studied. Bartelmez very generously invited me to spend a period there examining some of these embryos from this point of view. Unfortunately, I was not able to shunt my itinerary around sufficiently to spare a week or fortnight for such a study. Just eight or nine years previously my mentor at Wits, Joe Gillman, had worked on this Carnegie collection, to establish the earliest stages in the embryonic development of the human gonad. As embryos, we all go through a neuter-looking stage before the gender has microscopically declared itself as male or female. Gillman then studied the first declaration of maleness as the 'neuter' gonad transformed into a testis; or the first expression of femaleness in the metamorphosis of the asexual gonad into an ovary. Gillman's study was a classical contribution to the embryology of the sex-glands. For decades, his 1948 researches were quoted in embryology textbooks.

IT TOOK TEN hours to fly from Washington, DC to San Francisco, with a refuelling stop at Denver, Colorado. We came in over the Bay under a very bright, almost full moon, our shadow on the water growing ever larger and alarmingly so, till it had seemed we must ditch our aircraft – but the runway started on the shore. Professor Theodore ('Ted') McCown was there to meet me when I landed at 10.05 p.m. He drove me across the eight-mile Bay Bridge to Berkeley and installed me at the Durant Hotel close to the campus. I spent a week in Berkeley, interrupted by a day at the medical school in San Francisco.

Berkeley had been all but destroyed in a fire that swept down from the hills in 1926. When I was there, I saw many houses of redwood beams, breeze blocks, and a few stuccoed white exteriors reminiscent of Spanish architecture. It was the seat at that time of the largest campus of the University of California, with 16 000 to 17 000 students in a town of 120 000 people. The campus covered a large area and was overshadowed by a giant cyclotron high on the hills above, alongside the Atomic Energy Commission's building.

Ted McCown agreed fully with me that physical anthropologists need a grounding in anatomy. At Berkeley the Anatomy Department did not have space enough to give courses to undergraduates in anthropology, but graduates in physical anthropology were instructed to take anatomy. A special course was arranged for them in the Anatomy Department. You could not get a Ph.D. in physical anthropology without having done some anatomy. These practices were at least an approximation towards the system on which I was reared and in which I believed strongly. Earnest Albert Hooton at Harvard likewise sent his graduate students over to anatomy. Ales Hrdlicka, at the Smithsonian Institution, Washington, DC, went further: he used to say that it was necessary to be *medically qualified* to become a physical anthropologist and his protégé T. Dale Stewart followed his advice.

McCown was a veritable encyclopaedia of information about the development of physical anthropology in the USA. It was interesting to learn from him something of the academic dynasties. Almost every American physical anthropologist 50 years ago had been trained by Earnest Hooton of Harvard, or by protégés of Hooton. McCown was an exception, having been largely self-taught, added to which he had sat at the feet of Arthur Keith for four years.

Hooton had been very influenced by the work on races of another Harvard anthropologist: Roland Dixon. The earlier work of Dixon had a strong racist tinge and Hooton never escaped that slant – as illustrated in his work on the Old Americans, criminals, and so on. McCown added that, although Hooton had been a superb teacher, he had never established a 'school', in the sense of a school of thought, and hence his students had diverged into a wide variety of directions.

I have observed that leadership in science is often thought to connote the establishment of a school of thought. But some very great teachers have fired their pupils with the spirit of research, without necessarily pointing them in a specific direction or moulding in them a special mode of thought.

In my own case, when I became head of the Wits Department of Anatomy (see Chapter 25), it was a conscious policy decision on my part not to pursue any particular 'party line'. I wanted to have several modalities of interest and research activity enshrined within the department – including neuro-anatomy, cytogenetics and human genetics, embryology – descriptive and experimental, physical anthropology of the living peoples of Africa and palaeo-anthropology of the long-dead, growth and developmental anatomy, macroscopic and microscopic anatomy, teratology (congenital malformations and their development), and so on. The teaching offered reflected this diversity, and so did the research work of the department. The topics chosen by my Master of Science and Doctor of Philosophy students, numbering over 50 during my 53 years' sojourn in the department similarly addressed a diverse array of topics, although there were some areas in which efforts were clustered.

While I was with Ted McCown he nominated me for membership of the American Association of Physical Anthropologists (AAPA). My seconder was Larry Angel of Jefferson Medical College in Philadelphia, who was the secretary-general of the Association. This was accepted and in that very year, 1956, my name appeared in the list of members and I presented my first paper at its conference in Chicago (see Chapter 19).

This was one of my earliest memberships of an international scientific organisation and I have been a member continuously for almost 50 years. Despite its name, this organisation is the most effectively international body catering for the needs and interests of physical anthropologists in every corner of the globe. Its cosmopolitanism is testified to by the geographical dispersal of the authors contributing articles to its periodical, the *American Journal of Physical Anthropology*. The other international memberships to which I was elected in the 1955–56 period were the Anatomical Society of Great Britain and Ireland and the Royal Anthropological Institute of Great Britain and Ireland, both of which are also effectively international bodies.

A SUCCESSION OF two buses carried me southwards from Oakland, one of the 'East Bay Cities', to Palo Alto. There I was met by Kathleen Gough, whose husband, Dave Aberle, I had met at Cambridge. While she had carried out four years of fieldwork in India, he had worked mainly among Navaho Indians. On my first day in Stanford – 2 February 1956 – I attended a seminar on the biological basis of language and participated in the discussion. This was to be a major interest of mine from the 1970s onwards.

In 1956, Stanford had about 8 000 students, was rather select, and full of tall, long-legged, loose-limbed striplings. The sight of them intensified my thinking on the secular trend among modern people, especially in First World communities, and illustrated a marked increase in stature under Californian conditions.

The university consisted of a number of smallish, hand-quarried sandstone buildings, quite charming and unpretentious, scattered amid parkland, abundant native oak trees and a rich profusion of palms and other introduced species. The Ford Center for the Study of the Behavioural Sciences, which had started functioning in September 1954, was high on the tree-covered hills above the campus. About 52 people gathered there to 'think' – it was nicknamed the Ford Think Center, and even the Ford Think Tank. There were no strings: Fellows were free to write up accumulated unpublished material, to read, talk, confer, write books, with no administrative or teaching responsibilities. They were paid a university salary plus 10 per cent. It was open to people from other countries, but up to the time of my visit only one Englishman and two Israelis had so far come there. Two ex-South Africans (both of whom started out in the Wits Anatomy Department) subsequently played a big part there, the late Maxwell Cowan and the Nobel laureate Sydney Brenner. Forty-eight years ago, when I first became aware of the 'Ford Think Tank', and visited the premises, I contemplated the possibility of some day applying to spend a year or so there: I had, and still have, so much unpublished material and books mooted or requested that it seemed to me to be a marvellous prospect and brilliantly conceived by the Ford Foundation. But the passage of the years and decades has never allowed me a block of time away from home, such as might do justice to the purposes of the Ford Center. It remains a vision among my file of 'if only' and 'might have been' dreams.

The head of the Department of Anatomy at Stanford was Professor William ('Bill') Greulich, a man with whom – as with so many other new acquaintances in the USA – I was to develop a lifelong friendship.

Greulich confirmed my impression that the campus was full of tall people. In one year recently, he said, over 60 per cent of first-year students were 6 foot (1.8 metres) or over.

We discussed how already anatomy had become unpopular in the USA. Greulich reflected that part of the reason for this was the olden-time approach to the subject: anatomy had long been looked to as a means to weed out inadequate medical students. As a result, it had been a big and unpleasant hurdle to get over. It is said that some anatomists revelled in getting students to identify carpal bones (those awkward, tricky, little wrist-bones) by touch with their eyes closed. No wonder many young, impressionable students had received a raw deal at the hands of anatomists. This had turned them against the subject, which reflected itself in professional and faculty attitudes. The professor of one of the clinical subjects recalled how he had never forgotten being described as a 'butcher' by a former professor of anatomy, and how deeply it had affected the class when the professor had said, 'There's an abattoir down the street'. It was all part of the aura with which anatomy had become endued. This, coupled with the need for more time for 'new' subjects such as biochemistry, led to the whittling down of anatomy as a curricular subject.

It was discussions such as these, as well as some of my own experiences in the Wits Anatomy Department, that led me – when I was appointed to the chair in 1959 – to formulate a contrasting philosophy of anatomy without tears.

16 Los Angeles and the Grand Canyon

The train journey from Palo Alto to Los Angeles was memorable. After leaving Palo Alto on a Southern Pacific train, my coach connected at San Jose with the *Coast Daylight* from San Francisco. I travelled 'coach' (the term was new to me), that is, in a seat in a large unpartitioned coach. The journey was smooth, silent and comfortable. We traversed the famous fruit-growing valleys between the coastal range and another parallel range. In the afternoon, we emerged from the coastal mountains, after winding down a spectacular mountain railroad to San Luis Obispo. For over a hundred miles, from there southwards, our track lay along the coast, sometimes almost on the beach with a little sea spray reaching our windows, then on the dunes, but most of the time in full view of the Pacific Ocean. We drew in to the station at Los Angeles at 6.00 p.m.

My cousin Betty Mordecai and her husband, Dave Zeldin, who lived in Hollywood, met me at the station and installed me in my hotel, the Alexandria, in downtown LA. It was not far from 'Skid Row', where, I was warned, you were liable to be knocked over the head for a dollar, or for the very clothes on your back. I was warned not to go into certain streets at night. However, the proofs of my chromosome book were catching up with me from London and, one night, having finished correcting a set of proofs, I felt I had to post the batch. There was no postbox in my hotel. I asked the desk clerk where the nearest postbox was and he gestured – that way, right in the midst of 'Skid Row', a few blocks from the Alexandria. Against his advice that on no account should I walk down alone at that time of night, I set out with my package. The surroundings became murkier and more ill-lit with each passing block. I walked briskly in the middle of the road – fortunately there was no passing traffic – found the postbox, deposited the package, and walked back even more briskly to the Alexandria. My parcel reached London and I reached my hotel safely, but

I did not again test the midnight mysteries and potential perils of LA's 'Skid Row'.

If there was no direct threat from the citizens of the street, there was one from the tectonic instability of the area. When I arrived in LA, I found that the city had no really tall buildings. There was a limit of twelve floors with only the new City Hall allowed to exceed this. The reason, of course, was that this whole southern California region had relatively frequent earth tremors. In fact, at 6.17 p.m., seventeen minutes after I arrived in the city, there was a rather strong tremor in the LA district, followed by another tremor one-and-a-half hours later. But it was on 9 February 1956, at about 6.35 a.m., that I was awakened in my bedroom on the 11th floor by an especially severe tremor (measuring 6.8 on the Richter scale). At that height, I perceived the effects of the tremor as a remarkable swaying of the whole building and a rattling of the venetian blinds. It startled me and continued for fully 30 seconds, if not more. According to the papers, it was felt all over southern California.

It was a special privilege to spend time with Dr Joseph B. Birdsell, an associate professor in the UCLA Department of Anthropology and Sociology. He was about 42 then, with a slight chin-beard, and a drawl in his speech – apparently Harvard superimposed on a Middle Western (Indiana) origin. Birdsell was a student of Hooton's and he spoke of him with adulation.

Hooton had had great blows in his life. His two sons were both 'subnormal', as people were classified in those days. The third child, a daughter, was apparently normally developed. She married, but her marriage broke up and then she had a psychotic breakdown and had to be committed. The passion with which Earnest Hooton had thrown himself into studies on 'dysgenics' had to be seen in the light of this family history. When I learned of this tragic family record, I could not but remember Raymond Dart and his porencephalic son, Galen, and Dart's passion for better understanding of muscle control, posture and poise.

Birdsell had made two major expeditions to Australia. He had visited and seen almost all the surviving Aborigines (which he said numbered about 30 000). I was struck by the similarity in the numbers of these surviving and transforming hunter-gatherers, to the numbers of the San as I had determined them – and my article on these San numbers appeared that very year, 1956, in the journal *Africa*. Birdsell's list of data to be collected on each individual was based on Hooton's plan, to which Birdsell had added a number of genetical and anatomical traits. At that stage he had not published the complete report, but several papers had appeared. Birdsell had come to the conclusion that three main 'strains' had gone into the making of the Australian and Tasmanian people. These divisions were based essentially on morphological features, but they seemed to have some support from blood groups and serum proteins. However, all were recognisable as Australians who constituted a gene-pool; within it certain types could be distinguished, whilst many intermediates were of course encountered.

I could not help being struck by a resemblance between Birdsell's Australian analyses, and those of Dart and his school on the Khoe-San peoples. The South Africans were classified by Dart into 'types' labelled 'Bush', 'Boskop', 'Kakamas', 'Australoid' and so on. This was the diet on which I had been nurtured and initially supported and developed, but that Ronald Singer had overthrown. However, I think Birdsell's study was based on sounder human biological principles than were those of the South Africans working on the Khoe-San peoples. In any event, both approaches were essentially typological. This kind of analysis into 'types' has been largely overthrown by modern genetical analyses.

NOBODY WHO MAKES an academic visit to LA should fail to see some of the sights of the city. With the generous help of Joe and Rüsslein Birdsell, Betty and Dave Zeldin and their friend Brendan Delaney, I paid lightning visits to the Planetarium, high on Mount Hollywood, with an exceptional view of LA and the San Fernando Valley; Sunset Strip on Sunset Boulevard, with Wil Wright's famed ice-cream bar; Grauman's Chinese Theatre; beautiful homes in Beverly Hills; the coastline northwards from LA through Malibu beach to Zuma beach to watch a pair of grey whales spouting down the coast and there to sunbathe and picnic with Joe and Rüsslein.

There was a visit to the new Mormon Temple in LA, an amazing phenomenon in the middle of the twentieth century. Half a million people had passed through before it was closed to non-Mormons for dedication – shortly before I arrived.

Brendan Delaney took me to the Mexican Quarter where I was introduced to that fiery Mexican potion, *tachele* (tequila) – but I am afraid I have little recollection of the rest of that evening!

My hosts felt I could not omit Forest Lawn Memorial Park. This fantastic burial ground had provided Evelyn Waugh with material for his devastating satire, *The Loved One: An Anglo-American Tragedy*. It was laid out in spacious, hilly gardens amid trees and marble statues, replicas of classical masterpieces such as Michelangelo's *David*, and a stained window based on Leonardo's sketches for *The Last Supper* by Morelli, who we were told was one of the last members of a family in Italy who had handed down the secret of glass-staining from generation to generation – a technically remarkable piece of work. I hope I shall not lose any of my Los Angelino friends if I quote what I was moved to write in my journal after the visit: 'Forest Lawn is, of course, an extreme example of contemporary American culture: but I feel it did not really need an Evelyn Waugh to bring out its funny points – for it is a satire in itself.'

I should not forget to mention the Hollywood Farmers' Market. I marvelled at the *soigné* displays of food. Most intriguing was the counter of what were not yet being called 'ethnic foods': grasshoppers (fried, salt added), tinned rattlesnake with supreme sauce, tinned cuttlefish in its ink, plain, boiled quail eggs, broiled octopus on skewer, fried worms (from Mexico), smoked frogs' legs, tinned bird's nest soup, whole quails, snails. Everything was spotlessly clean and, although I did not have a good appetite, I relished the sight of the taxonomic array of foodstuffs. It was many years later before I had opportunities to visit Mexico, Costa Rica, Ecuador and Brazil, or Japan, China, Vietnam, Malaysia, Thailand and Indonesia – and to savour an even greater variety of eatables that the populace ekes out of its teeming ambience.

ON 12 FEBRUARY 1956, as a fog rolled in over LA, I left the Pacific behind and started towards the east. I was aboard the Acheson, Topeka and Santa Fé Railroad, the Grand Canyon Limited. The scenery was beautiful as we climbed up the mountains to over 7 000 feet (2 134 metres). During the night we crossed from California to Arizona, awaking in the Pullman car on the station at Williams, Arizona. We drove by bus through lovely, snow-dusted pine forest to the Grand Canyon. All day Monday, I admired the Canyon from different vantage points along the west and east rims. The meandering of the Colorado River in the floor of the Canyon extends for 350 kilometres. The average depth of the Canyon is about 1.6 kilometres and the width across some 16 kilometres. It was amazing to contemplate that at about the end of the Mesozoic (the geological time marked by the development of dinosaurs, and with evidence of the first mammals, birds and flowering plants), the land level was low and there was a sluggish Colorado River. Then the lands began to rise and the river to cut down. The present rim is the Tertiary surface, the Quaternary deposits having been eroded above our heads. This surface is 2 350 metres above sea level. At this level seashells are found! There are seven successive layers of limestone, indicating seven successive inundations by inland seas. There is a Great Unconformity where Tertiary deposits rest on Primary. The basal black rocks, we were told, have no evidence of life. The next layers have fossil algae; the next – animal life (Trilobites); and so on. The Canyon has been a great ecological barrier. For instance, the squirrels and gophers are different on the two sides. The northern rim is 365 to 457 metres higher than the southern edge; and to the north there is a Canadian type of forest, instead of the pines and junipers of the south. The north has a mean annual snowfall of 510 centimetres, whilst on the south it is only 190.5 centimetres. Six out of seven of the world's major climatic zones are represented, from near-desert conditions at the bottom to the sub-arctic tundra conditions of the San Francisco Mountains, 80 or 90 kilometres away. Only the tropical humid zone is not represented.

The rocks are mainly sandstone, limestone and shale. At the head of the Canyon is the confluence of the Colorado and Little Colorado Rivers, and there is a major north-south fault at the end. Probably both factors help to explain why the Canyon is situated where it is.

Overshadowing all the scientific facts about the Canyon, the sheer beauty and enchantment of this natural wonder filled my very being with rapture. As I looked and strove to understand, the knowledge of the Canyon's evolution mingled with the sensation of pleasure and ecstasy: each played into the other's hands. It was a very telling experience of the interface of knowledge and bliss. After all, as John Keats taught us:

> 'Beauty is truth, truth beauty,' – that is all
> Ye know on earth, and all ye need to know.

After a day of exaltation at the Canyon, we took the bus back to Williams and then another bus to Flagstaff, 51 kilometres away, where I checked into the Monte Vista Hotel. Soon after my arrival I sallied forth into the hotel bar thinking to have a nightcap and turn in. As things transpired, it was 7.30 a.m., next morning, before I got back to my room! I met some interesting characters in the bar, an air force colonel, someone from the local radio station (who tried to persuade me to record an interview the next day), and a recruiting sergeant in the Marines. They talked me into accompanying them to visit a ranch, a real one, and not a 'dude ranch'. Off we drove there and then, over 48 kilometres south through Oak Creek Canyon. There was a big party at the ranch. Some of the girls were frying eggs – but whether these were for supper or for an early breakfast I do not recollect. As I had been a little more abstemious than my companions had, I drove myself most of the way along a narrow, tortuous, precipitous road. When, by early morning light on the return journey, I saw the hairpin bends, the sheer rock faces and the precipices I had traversed at 1.00 a.m., after a few drinks, in a strange vehicle, and driving for the first time on the right-hand side of the road, I nearly fainted.

17 Sojourn in St Louis

On my first visit to St Louis, I spent just on a fortnight, mainly in the Anatomy Department, where my mentor, Raymond Dart, had spent time as one of the first Rockefeller Fellows 35 years earlier. I was installed at the Forest Park Hotel, about five or six blocks from Washington University Medical School.

The head of the Department of Anatomy was Ed Dempsey. I found him to be friendly and helpful and he gave me a full day discussing both research matters and Anatomy Department affairs. His interests had ranged over endocrinology, neurophysiology and, especially, electron microscopy, of which he was one of the pioneering exponents. He introduced me to the marvels of this high-powered tool. The fixatives used up to that time were suitable for studies on the cytoplasm and a unifying cytoplasmic hypothesis was emerging as a master concept. Within its purview, it could explain membranes of the cytoplasm and the nucleus, the Golgi apparatus, mitochondria, vacuoles, infoldings and outfoldings, basophilia, ergastoplasm, phagocytotic appearances, secretory granules and more besides.

It made me, as a chromosome, chromatin and nucleolus person, eager to see how the nucleus would look if a satisfactory nuclear fixative were available, of which there was none at that time. As a result of this lack of a suitable nuclear fixative, the nucleus had been largely ignored by electron microscopists. It set me dreaming that, if ever I were in charge of an anatomy department one day, the acquisition of an electron microscope would be high on my wish-list (although the latter term was not yet in use 50 years ago).

Ed Dempsey had strong views on the place and the role of lectures in the undergraduate course in his department. He personally was more interested in concepts and general principles, in terms of which the students could *understand* their anatomy. He conceded that some parts of the course needed didactic lectures and a fair amount of lecturing was included in histology (the study of organs, tissues and cells under the microscope). But the teacher, in putting over general concepts, had to be free to roam far afield in selecting examples, to illustrate ideas. He or she must not feel confined to the limits of only gross anatomy or only microscopical anatomy. This is where integration came into the picture.

'Integration' was already a term on many medical educators' lips at that time. Sometimes, it referred to horizontal integration between subjects and topics taught in the same year of study; often it was related to subjects taught in different years of study in the medical course, as vertical integration. For some it was a glib catchword, a sort of shibboleth, a piece of academic gamesmanship often thrown into the ring when the revision of curricula was under discussion. Dempsey's view, which I shared throughout my 32 years as head of the Wits Anatomy Department, was that the objective of integration should be integration in the student's mind. Mere synchronisation did not connote integration: it might not prevent integration coming about, but it would not per se bring about integration, except in the minds of those few students who were integrated anyway. The teachers must themselves be integrated. In lecturing, Dempsey felt free to make any excursion he pleased into physiology, medicine, experimentation. The degree of integration effected in the minds of students was a function of the teacher's outlook. Excursions into history, for example, may be helpful. The idea was not to restrict a lecturer, but to encourage him or her to influence the students as much as possible, based on the lecturer's body of experience. When 'authorities' differ – and the luckless student is moved to say, for instance, 'but Dr— said something different' – the differing authorities should gather round and thrash the difficulty out. Even the professor should be big enough to climb down if occasion demanded! It sets a good example, too: the student would learn that interpretations may differ; authorities may be wrong.

Dempsey and I had an interesting discussion on the place of physical anthropology in anatomy departments. In the Commonwealth countries, there was a long tradition of physical anthropology being carried out within anatomy departments: this applied within the UK, South Africa, Australia and New Zealand. In those countries relatively little of physical anthropology was pursued outside of anatomy departments. By contrast, in the USA, physical anthropology had long been closely associated with other branches of anthropology, such as social and cultural anthropology, linguistics and archaeology. Very few anatomy departments had physical anthropologists as members.

I was especially interested to see whether physical anthropology flourishes better under the 'British system' or under the 'American plan'. One important outcome of my six months in the USA, and I was especially bent on studying this question, was that the well-being of physical anthropology did not depend on the structural umbrella under which the subject reposed. Outstanding work in the various branches of physical anthropology was being carried out under both kinds of departmental organisations, and indeed others, such as in museums. Under the latter umbrella, admirable work was being carried out in institutions such as the American Museum of Natural History, the Natural History Museum (London), the Musée de l'Homme, Paris, and the Transvaal Museum, Pretoria. In broad summary on this question, as in so many other branches of education and research, whatever is, is right, as Alexander Pope said centuries ago.

ROBERT JAMES TERRY, the emeritus professor of anatomy at St Louis, then in his 86th year, was sound of faculties and had a keen wit. He was doing such incredible things as cannulating lung alveoli and drawing off the practically monomolecular film of fluid on the surfaces of the alveoli – for chemical analysis! Yet he had a glass eye on the right, had had a left-side cataract and was intensely photophobic. He was not beyond describing some ithyphallic people (with semi-erect penis) as 'half cocked'. Anatomists do have some very amusing and often outrageous stories!

Terry told me how it came about that he had started the 'Terry Collection' of skeletons of black and white Americans. He had studied at Edinburgh in the 1890s under Sir William Turner, who had built up a collection of human skeletons at Edinburgh. Turner had been the leading British anatomist of his day. He was an early pioneer in the brokerage of the marriage between anatomy and physical anthropology in the UK. After he realised the value of such a collection of cadaver-derived comparative human skeletons of modern mankind, the idea was transplanted to the New World. Carl August Hamann (1868–1930), in the Anatomy Department at Western Reserve University, Cleveland, Ohio, started building up a similar collection of human skeletons before the end of the nineteenth century. After 1912, his successor, Thomas Wingate Todd (1885–1938) added appreciably to the materials that later came to be called the Hamann-Todd Collection.

Raymond Dart visited St Louis in 1921 as one of the first two Rockefeller Foundation Fellows. Dart, in turn, carried the concept to South Africa, from which was built up at Wits an Old World collection of cadaver-derived human skeletons. Much later, as Dart's successor, I named it the Raymond Dart Collection of Human Skeletons. During my stewardship as head, I trebled its previous size. Thus, in the English-speaking world, at least, you can trace the genesis and dispersal of this type of human skeletal collection from the Old World to the New and then back to the Old World again. From Johannesburg, further offshoots of the concept went to several other parts of Africa and to Australia (Perth). Studies on such collections of human skeletons have been made in all parts of the world and continue to receive attention. Researches on human skeletal collections have yielded, and continue to yield, results of immense value to anthropology, orthopaedics, dentistry, developmental anatomy and human variation.

Terry told me that he was very worried about what would happen to the skeletal collection after he and Mildred Trotter (who was in charge of gross anatomy at St Louis) had passed on. He was trying to impress upon the university that, however much fashions might change, such a collection must always be preserved as a valuable source of research materials.

Mildred Trotter ('Trot') gave me a detailed insight into the objectives and arrangements of the anatomy course. She told me that the number of hours spent by the medical students on gross anatomy had dropped from 600 to 300 hours. I could not help reflecting on how dispiriting that must be for dedicated teachers of gross anatomy – and Trot was one such. In her course there was a strong emphasis on observation: the student had to learn to use her or his eyes in studying the cadaver. That is why Trot discouraged books such as *Cunningham's Manual of Dissection*. While Trot's philosophy was to urge the students to 'see what the body says', the book tells them what to look for and how it looks – and therefore actively discourages observation. I have always held the same view and often pointed out to my clinical brethren that cadaver anatomy, if properly taught, was the only part of the entire medical course in which students were encouraged to develop their powers of observation.

Although it was clear that the dissection of the cadaver was the best way to impart a detailed knowledge of anatomy to students of medicine and the allied health sciences, cadaver anatomy was really only a beginning. It was patient anatomy that we needed to get across, for the students' patients were for the most part alive and although, when very ill, they might be cadaverous, they were not cadaveric! Indeed, teachers of anatomy would do well to remember that the examination of a patient classically starts with inspection and palpation, that is, observation with the eyes and with the fingertips.

What means did we have to bridge the gap between cadaver anatomy and patient anatomy? There were two recognised – though in my view underused – approaches: X-rays including scanning radiographs (a quite marvellous recent innovation), and living anatomy, comprising surface and functional anatomy.

In our Wits department, my fellow gross anatomist, Dr Maurice ('Toby') Arnold and I ran a weekly programme for about 20 to 25 weeks. Once a week, students were requested to bring bathing attire and to strip down in the dissection halls. For an hour they traced on the surfaces of one another's bodies the position and outline of organs and joints and other structures below the surface. Each week the region chosen was the area that had been dissected that week. Some of the living anatomy practical classes were functional in character: students were challenged to work out the anatomy of sitting, standing, walking, running, leaping and, to dip our wing to South Africa as a sports-mad country, the anatomy of kicking, bowling, batting and swimming (over-arm, butterfly and backstroke). To this blend of surface and functional anatomy we gave the name *Living Anatomy*. It was a significant part of our medical course in anatomy from the time when Dart put myself and Toby Arnold in charge of the gross anatomy course, at least until the end of my term as head of the department, that is, for at least 40 years.

Although we had, I believe, every reason to be proud of our course in living anatomy, we were not so advanced in our use of the X-ray machine as an aid to the teaching of gross anatomy.

When I read an article on the role of X-rays in the teaching of gross anatomy by a certain Wayne A. Simril of St Louis, I received his advice with enthusiasm and entered into correspondence with him. During my visit to St Louis in 1956, I arranged to call on him and he showed me his excellent set-up. There had been a scheme for collaboration between the Departments of Anatomy and Radiology, but this broke down with a change in the headship of the Department of Radiology. We discussed what he and I from different standpoints considered to be an ideal arrangement and worked out a two-stage programme to meet two objectives. In Stage 1, the course in radiological anatomy, an anatomy department would make use of radiology and radiologists as an aid to the teaching of anatomy and, especially, to help bridge the gap between cadaver anatomy and the patient's body. In Stage 2, the course in anatomical radiology, the emphasis would be on the needs of the radiologist, first on the anatomy of the normal, later of the abnormal. In this arrangement, the co-operation of the two species of medical scientist – anatomist and radiologist – was essential.

MILDRED TROTTER BECAME one of the leading authorities in what has come to be called skeletal biology. She and I kept close contact with each other for decades. Years later I was invited to deliver the 29th Robert J. Terry Lecture in the Washington University Anatomy Department. It was 1981 and I was in my second year as dean of the Wits Faculty of Medicine. Although my duties in the deanery led me to decline most of the invitations I received from 1980 to 1982, in the case of this invitation – from the chair of anatomy, Roy Peterson, and from Trot – I felt I could not possibly say no. After all, Wits and Washington University (known widely as 'Wash U.') had had a long association going back to the early years of the twentieth century.

In the 1980s, through an initiative of Glenn Conroy of the Wash U. Anatomy Department, and of myself at Wits, together with Yusuf Dinath, the manager of the Wits Medical School, a system of exchange studentships was started. Senior Wits medical students, in their clinical years of study, were selected to spend an elective period at St Louis, while in the reverse direction Wash U. clinical students were selected to spend a corresponding period at Wits and its teaching hospitals. The exchange scheme was generously supported by funds raised by Conroy in St Louis and it lasted for a number of years in the latter part of the twentieth century. This further cemented almost a century of valuable collegial relations between the two institutions in two hemispheres and two continents.

In 1984, my colleagues and I organised an International Congress at Wits and at the University of Bophuthatswana (later the University of the North West) in Mmabatho to celebrate the 60th anniversary of the discovery of the Taung skull. Three United States octogenarians attended: Mildred Trotter of St Louis, T. Dale Stewart of Maryland and formerly of the Smithsonian Institution in Washington, DC, and W. Montague ('Monty') Cobb of Washington, DC (see Chapter 20). It was a special privilege for the other participants and for my South African friends and colleagues to meet these three veterans.

It was very hot at Mmabatho. Trot was clearly not feeling herself. Her friend from St Louis, Julia Otto, and I took her from the crowded auditorium to sit on a bench outside in the shade. Suddenly, I saw Trot go pale and she keeled over unconscious on the bench beside Julia and me. Fortunately, my dearest friend and general practitioner, Dr Joseph ('Joe') Teeger, was attending the congress to keep an eye on the health of the participants. I rushed in to fetch him from the auditorium. Joe confirmed that Trot had had a stroke and would need to be removed to a hospital. In no time at all, he was on the phone to the Johannesburg Hospital adjacent to the Wits Medical School. A bed was booked for Trot and the Emergency Helicopter Service was asked to send a 'chopper' to the University at Mmabatho, about 400 to 450 kilometres away from Johannesburg. In the meantime, we carried Trot to a comfortable bed and waited for the arrival of the helicopter. She had indeed sustained a stroke and had lost the power of speech. Trot spent about six weeks in the Johannesburg Hospital before she was deemed to be fit enough to fly home. During her recovery period in St Louis her ability to understand spoken language and to communicate gradually improved, but sadly she did not recover her speech, although she could sing. I kept in touch with her through Julia Otto and others and was greatly saddened by her death in 1991.

18 Human genetics in Ann Arbor

If I dwell on some of my train journeys as I criss-crossed the United States, I do it because I imagine few of my colleagues and readers today, including visitors to America, use any public transport other than aircraft. Just as I sailed by passenger liner from South Africa to Britain and between Britain and America, so I saw America by a thorough sampling of the railroads. To travel from St Louis to Ann Arbor, I had to go first to Chicago and then take another train to Ann Arbor. I left St Louis on the *Wabash Domeliner, Blue Bird*. This was fitted with Vista-dome cars, in one of which I was ensconced (another first for me). It was a pleasant experience sitting up there and watching the countryside flash past, drab, dun-coloured, flat. We crossed the mighty Mississippi River about eleven kilometres from St Louis. Arriving in Chicago at 2.00 p.m., I had to change from Dearborn to Illinois Central Station. There I boarded the New York Central's *Twilight Limited* for Ann Arbor. It was cold in Chicago – my furthest point north so far. Increasingly the ground was ice-covered and snowbound as we approached Ann Arbor.

I nearly missed the station. It was 8.30 p.m. by my watch and we were due in at 9.15 p.m. I could not fathom why we were, as I thought, forty-five minutes early. I had not been warned that between Chicago and Ann Arbor we crossed a time-line and it was really 9.30 p.m. – we were fifteen minutes late. I had a wild scramble from the washroom, gathering all my belongings, and alighted in the nick of time.

Professor Bradley Patten was waiting for me. He drove me over ice-covered roads and installed me in his home. I was received by Barbara Patten whom I had last seen in bed with a bad cold in the Carlton Hotel, Johannesburg, in February 1954.

The Patten house was on the outskirts of Ann Arbor, close to wooded hills that were still white with snow, though thawing freely. The garden and surroundings were alive with birds and Barbara helped me to identify chickadees, cardinals and blue-jays. Some beautiful pheasants, of which I saw one or two males and a harem of females, used to feed early in the morning on corn-cobs set out in the garden for them. Fox squirrels, red squirrels and rabbits were other early morning visitors to the garden. Barbara was a keen amateur ornithologist and did bird-banding for the national bird-banding service run from Washington, DC. There were several traps in the garden and on window ledges and Barbara pulled a string inside the house to release the doors whenever she saw a bird inside eating the seed-bait. The bird's leg was then marked with a band. On Sunday, 4 March, we spotted a squirrel in one of the bird-traps. The Pattens caught it in the cage and, as was their practice, drove away with it in the cage for many miles before releasing it.

After we had dropped the squirrel from the cage, we had a pleasant afternoon's drive to a recreation park about 45 kilometres from Ann Arbor. The river was partly frozen and I took a walk out over the 15-centimetre-thick ice. It was another entirely new experience for this son of the sub-Tropics. On the unfrozen part of the river there was a flock of Canadian geese. People in little shacks were ice-fishing through holes in the ice. Others were skating although it was late in the season and most of the winter sports had stopped. It was cold and sunny, but not too early for us to see and hear a group of meadow-larks.

THE ANATOMY COURSE at the University of Michigan was longer, and more integrated into the medical curriculum, than in any other American school I had visited. The department was the nearest to ours at Wits that I had found, in terms of the amount of teaching and the multiplicity of courses offered.

Dr Russell T. ('Russ') Woodburne, the professor of gross anatomy, ushered me into the minutiae of the gross anatomy course. His research interests earlier were neuro-anatomical and then became more focused on the vascular system. When I visited, he had become 'just a gross anatomist'. He was writing a textbook in which the material was arranged on a regional basis, because the dissection went along regionally. The students found it difficult to use a text based on the *systems* of the body, like *Gray* or *Cunningham* or *Morris* – which in any event, Russ said, were too detailed.

His new book would combine the features of regional arrangements and less detail (though more than Grant's *Method of Anatomy*, which, he held, had insufficient detail).

Years later, at Wits, it was because I had been similarly dissatisfied with the existing texts, that I, at first with surgeon-anatomist Toby Arnold, and later, with surgeon-turned-anatomist, John Cameron ('Jack') Allan, had embarked on our own textbook. On the advice of the well-known South African writer and novelist, Sarah Gertrude Millin, we were to call it *Man's Anatomy*. When the first edition of our book appeared in 1963, it was before the dawn of the era of political correctness. It was only when we were preparing the fourth edition, about 25 years later, that I received the first criticism of our 'politically incorrect' title from a female professor of anatomy at Honolulu.

To my amazement, I found that the anatomy course at Ann Arbor filled no fewer than 482 hours (excluding histology, organology and embryology). At that time the corresponding number of hours in the Wits Anatomy Department was about 420 hours. Since that time, anatomy courses in especially anglophone countries have followed a different fashion and such totals have dropped appreciably. This trend does not seem to apply to European, African and Indian departments, where the place of anatomy in the medical course is still large and respected.

I HAD NOT forgotten my juvenile vow to become the first medical geneticist in South Africa, after my puzzlement over my sister's death in 1942. By the early 1950s there were still no human heredity clinics or departments at any South African universities. I started to give a course in medical genetics to Wits psychiatry candidates at that time and, presently, to other aspirant medical specialists. As a result, a number of patients were referred to me for heredity counselling. The kind of problem that a family faced was, for example, that of a mother who had delivered a malformed infant, perhaps one with spina bifida or microcephaly. What were the chances for her next child? Should she have another baby? Fifty years ago the counsellor had to rely on the statistical probability, calculated from published family histories.

One of my primary reasons for coming to the University of Michigan in Ann Arbor was to study human heredity and counselling techniques at the knee of a leading genetics authority. Dr James V. ('Jim') Neel was the newly appointed head of the recently created Department of Human Genetics. Like myself, Neel held both Ph.D. and medical degrees. He told me that, as soon as he started going seriously into human and especially medical genetics, he found it insufficient to have a doctorate of philosophy in genetics and he embarked on a medical course as well. As a result, he was thoroughly at home in the wards and, in fact, still did ward rounds and had routine hospital duties. Thus, he was a sort of hybrid, having feet in both camps, the clinical and the scientific. As such, he served as an intermediary between the hospital and the Human Genetics Department. He could talk the language of the clinicians to them and then interpret or translate into genetical concepts and jargon.

Neel and his departmental colleague Jack Schull had played an important part in trying to determine what had happened to survivors of the Hiroshima and Nagasaki bombs. The American investigative team, led by Neel, worked closely with Japanese scientists in this unique follow-up to a gruesome episode in the history of war and science. At the time of my visit in 1956, Neel and Schull were preparing a book on the results of their researches. This appeared the same year under the title *The Effect of Exposure to the Atomic Bombs on Pregnancy Termination in Hiroshima and Nagasaki*. At that stage they had found little evidence of an increased incidence of malformed babies, which had been expected. There were some hints of a change in sex-ratio (the ratio of the number of boys born alive to that of girls born alive), and in the survival of live-born infants in relation to the radiation history of the parents. Although the major clinical programme had been terminated in early 1954, the collection of data on these two borderline findings was continued.

The human geneticists, as I have said, worked closely with the Ann Arbor clinicians and rooms were maintained in the hospital for the taking of genetic family histories. Whenever a case was referred, one of the geneticists on duty went to the ward to take a family history. Doctors in training were taught to take a medical case history of each new patient on admission, but a family history for genetic purposes is a very different matter and data are sought on every relative, whether or not he or she is ill or shows the same genetic condition as the original patient. I sat in on several such sessions and learned a huge amount.

Subsequently, Neel and his team undertook population genetical studies on hunters and gatherers of the Yanomama tribe in Brazil. Collectively these studies constituted one of his greatest contributions. There was a certain parallel with the work of the Wits Kalahari Research Committee on the San (Bushmen) of the Kalahari. We had discussed our studies on the San when I was in Ann Arbor and I had lectured on some of our results. Over the years, whenever the occasion presented itself, Jim and I discussed his researches on the Brazilian hunter-gatherers and ours on the living peoples of Africa, such as the hunters and gatherers of the Kalahari, the San, and the pastoralists and agriculturalists, such as the Tonga of southern Zambia. There was broad agreement on the principles underlying both the African and South American studies. I think, however, that Neel's Brazilian studies included a greater clinical component, while ours had a stronger morphological slant.

On my return to Johannesburg, I reoccupied my place in the Wits Anatomy Department. Over and above other duties, I resumed unofficial tasks as an heredity counsellor. I still have the files I accumulated on the families who sought my advice. This duty continued until the Anatomy Department recruited a short, curly-haired Welshman, Dr Trefor Jenkins, as a lecturer in human genetics in 1963. He took over that function and when, in 1975, Wits decided jointly with the South African Institute for Medical Research to set up a Department of Human Genetics under Jenkins as the first professor, the function of heredity counselling naturally gravitated across to his department. This was thoroughly justified, as the subject was so important as to demand a formal departmental structure.

In 1981, I was invited to attend the Sixth International Congress of Human Genetics in Jerusalem. Jim Neel was the Congress president. Neel's presidential address to the Congress was devoted to 'The wonder of our presence here'. What gave him his feeling of awe and wonderment was:

> a sense that our genetic organisation and how it got that way is a far more complicated subject than seemed the case during the period of scientific euphoria that followed the elucidation of the structure of DNA and the cracking of the genetic code. A genetic organisation that has been evolving 3 to 4 billion years can scarcely be expected to have

> yielded to our assault in the short, 30-year period since these seminal developments. Perhaps the principles that govern this organisation will in retrospect seem as simple as those governing the code, but for now that thought can only be an article of faith.

Then Jim spoke of the

> pervasive sense of wonder at the many vectors that over the long evolutionary haul somehow converged to bring us here, that same sense of wonder experienced as a child in its rapidly expanding world. We are a privileged group at a privileged time in the development of scientific thought and understanding – privileged not in the arrogant use of power but to witness and contribute to an unfolding story whose final implications we can only dimly foresee.

Well spoken, Jim! On so many occasions I have preached parts of this gospel to my students, both the very young – and perhaps especially to them – and to my graduate and postgraduate students. 'Never lose your sense of wonderment,' I have repeatedly said. On the other side of the coin, how sad I have felt, and how sympathetic, when I have been confronted, happily not often, with a student who is blasé, uninterested, beset with a closed mind. Such people are a challenge to the teacher and to the idealist, and when both are combined, as in myself, when the mentor is brimming with an overwhelming sense of wonder, it is doubly challenging. When my path of life has brought me into contact with students of that ilk, my pity is swiftly transmogrified into greater personal effort, of both the spoken word and the even more telling silent example. Over the years, I have had some 10 000 students through my hands at the Wits Medical School. I think I have been tolerably successful with them. At least their feedback, from younger and older days, from home and abroad, has led me to be thankful that my efforts have not been in vain. The retention of my personal sense of wonderment and of enthusiasm has, I feel sure, played a big part throughout my life.

19 The whirlwind tour continues

The third and fourth months of my stay in the USA were filled with brief visits to many destinations. For example, I spent three and a bit days in Detroit, during which I had two evening parties, three luncheons, gave four lectures (on the San and the australopithecines), and had eight interviews. This was the busiest spell on my 1956 American tour, but there was a rather hectic 24-hour day at Los Angeles and Santa Barbara in 1968 that was a close competitor in the exertion stakes!

From Detroit I travelled to Chicago where the University of Chicago Department of Anthropology was one of the most impressive and active departments I visited in the USA. It had a team of high-powered people such as Sol Tax, Sherry L. Washburn – one of the most influential figures in American anthropology at that time and for many years afterwards – Robert Redfield, and the new appointee, Francis Clark Howell. There was a terrific team-spirit and progressive outlook among them. They had no fewer than 72 research students (M.Sc. and Ph.D.), the highest total in the USA. Of the 72, some 14 were in physical anthropology. That is to say, Sherry Washburn was training many of America's physical anthropologists of the morrow. This was excellent, as he had a valuable, dynamic, functional and anatomical approach to the subject. My main, or only, criticism was that he tended to neglect the historical and the non-American aspects (see Chapter 20).

In Chicago I delivered my first conference paper in the USA to the American Association of Physical Anthropologists (AAPA). A glance at the programme revealed rather interestingly the trends in the development of American physical anthropology: about one-third of the papers were of an *applied* character, human identification, air force anthropology and so on. Apart from one or two exceptions, there was an almost complete absence of genetics papers in the programme.

My paper on the evolution of the Bushmen was the only one of all those I heard in which an attempt was made to relate morphological traits to their genetic causation. Perhaps this still reflected the swing of the pendulum away from the racist anthropo-genetical ideas of the Germans. It was most unfortunate for anthropology that so few physical anthropologists at that time were using a genetical approach in their work. I stressed this point in my paper: we must get beyond blood groups and taste-testing, I pleaded, if ever our morphological analyses were going to become meaningful in genetic terms, that is, if ever we were going to understand how different varieties of human beings (not different populations of blood groups) had come about.

I suggested that the Bushmen (San) constituted a peculiarly African line of human evolution, which could be traced back to 'the Rhodesioid group', epitomised by the Kabwe or Broken Hill cranium from Zambia. At that time the skull in question, with its huge brow-ridges and deep, broad upper jaw or maxilla, was generally attributed to 'Rhodesian Man', hence my use at Chicago in 1956 of the term 'Rhodesioid group'. I indicated that if this type of human, which some were calling 'Neandertaloid' because it bore a superficial resemblance to Neandertalers in several features, such as the beetle-brows and large, inflated maxillae, had metamorphosed into such as the San, with small, smooth brows, petite jaws and teeth, it would have required genetic changes of two kinds: differentiative and size changes. The differentiative changes would have been from the gross, heavy, gerontomorphic features of the Rhodesioids to the delicately constructed, virtually infantile characteristics of the San. The size changes would have been from larger to smaller. I found reason to propose that the differentiative re-modelling occurred earlier and the dwarfing later. On this view these re-fashionings spawned the definitive Bushmen or San people. The thought seemed to excite a number of the participants. I remember, in particular, Carleton Coon (see Chapter 21) leaping up during discussion time, and declaring that if we accept that Rhodesian Man could have changed into Bushman in so short a time as the archaeological record indicated (and he used the rather wild 'guesstimate' of 50 000 years), then we would need to rethink and revise our very ideas on the tempo of human evolution!

Then, to replace William L. Straus of Johns Hopkins, who had been involved in a train accident en route to Chicago, I was asked at short notice to speak on the evolutionary development of the australopithecines. As I had been keen to bring in Dart's 'osteodontokeratic culture', I interpreted australopithecine ancestry very broadly, under four main heads: morphological, geographical, chronological and cultural ancestry. Under the latter head, I adduced about twelve lines of evidence, all supporting the idea that, if we confine ourselves to a single-cause analysis, it was most likely that australopithecines, and not hyenas, were responsible for the damaged bones found in such numbers at Makapansgat. But of course, not one but several factors might have produced the bone, tooth and horn fragments. I became most animated in my presentation and think I succeeded in rounding off the symposium satisfactorily – or, at least, so I was informed by a number of individuals.

It had proved to be an exciting and rewarding debut from a number of points of view, not least that, having been accepted as a member of the AAPA at this Chicago meeting, I had sung for my supper!

THE CHICAGO MUSEUM of Industry and Science was opposite my hotel. Everything that opens and shuts was there, opening and shutting. You could be conducted through a working model of a coalmine; you could pick up telephones in front of each exhibit and set off a running commentary on the exhibit; voices started explaining as you crossed the threshold of some rooms; lights flashed on, highlighting certain features of some exhibits. The medical section was outstanding: there was a remarkable series of human embryos, from the unfertilised egg up to the full-term infant (including a pair of identical twins at about three to four months of intra-uterine life). While I was inspecting this probably unique exhibit, an unescorted group of diminutive children came in and eagerly went along the succession of embryos. In front of each, they craned their heads, their eyes fixed on the middle region of the embryo – and I heard a piping of small voices: 'Little boy', 'Little girl', 'Can't tell'. Absolutely fascinated, as an anatomist and a student of human beings, I watched this tiny procession and then was rewarded to see that the eyes of adult viewers went first to the faces and then down to the genitals. Would somebody please call Dr Freud to elucidate these curious observations?

Close by was a huge model of a heart in which, to a loud lub . . . dub, lub . . . dub refrain, you could climb inside the right ventricle, through a door in its wall, inspect the anatomy from within, then cross through a patent inter-ventricular window, to the left ventricle and out again to the same lub . . . dub, lub . . . dub refrain. Another model of the human body responded to the touch of a button, by showing what happened to a single red blood corpuscle on a single beat of the heart: a little mannequin ran out, at about the speed of blood, and returned again from hand or foot. Yet another series of exhibits answered questions on heredity, heart disease, etc. Press the relevant button and the answer flashed out of a mirror. Here were visual aids carried to the nth degree. Such contrivances are commonplace today, but 50 years ago they were uniquely pioneering.

As I wandered around the Museum I pondered the fate of the voice as a means of communication in the future. Although I don't like to generalise, I had found that so many Americans presenting papers at conferences were barely articulate; sentence followed drab, over-constructed sentence, with nary a glimmer of relief, tone or humour. Few indeed recounted their contribution without a written paper in front of them. Sherry Washburn was a notable exception. He was fluently and mellifluously articulate. Was interaction with technology busy replacing the human touch?

My visit to the Chicago Natural History Museum yielded two highlights. The first was when Dwight Davis showed me the jaws and teeth of the giant panda. Although it was a member of the bear family of carnivores, it had become adapted to a diet of bamboo. The surfaces of its molar teeth were of immense size as though to enable it to masticate its tough fibrous diet. He drew attention to the possibility of a similar phenomenon in the robust australopithecines. In contrast with the teeth of *Australopithecus africanus*, those of the robust chaps, *Australopithecus robustus*, were greatly expanded and Davis suggested that this had evolved to enable the robust ape-men to cope with a more vegetal diet. Independently, John Robinson had relied on other features of the cheek-teeth of the robust ape-men to claim that they were adapted to a vegetarian diet. It seemed an interesting parallel. Years later biochemical analyses of samples of the ape-man teeth by Professor Nic van der Merwe and Dr Julie Lee-Thorpe of the University of Cape Town, tended to discount the hypothesis of Robinson and of Davis.

The other outstanding feature of my visit to the Chicago Museum was to pay my respects to the Man-Eaters of Tsavo, a pair of magnificent maneless lions that had between them been responsible for killing and eating about a hundred railway workers, when the line was being laid from the Kenyan coast to Nairobi. Lt. Col. Paterson had been in charge of the operation and eventually succeeded in shooting the lions. His book, *The Man Eaters of Tsavo*, gave a graphic account of these dramatic nocturnal tragedies. Dr Brian Paterson, the colonel's son, was an eminent palaeontologist whom I met on various occasions in Nairobi and at Harvard. He told me that the two lion skins had lain for years as floor-mats in a private home before they were identified through Colonel Patterson and given to the Chicago Museum.

MILWAUKEE, WISCONSIN, WAS a pleasant and easy drive from Chicago along the lake. The 69th meeting of the Association of Anatomists was held there from 4 to 6 April 1956. In my paper, I spoke on some of my doctoral work of a cytogenetical character, drawing attention to the nucleolar organisers, that is, loci or constant positions along chromosomes, opposite each of which a nucleolus is formed during the cell cycle. I proposed that discriminative studies on nucleoli, their formation and subsequent life-history, could throw light on the functional state of a cell. Also, they could furnish a useful tool for 'dissecting' the interphase nucleus, between cell divisions.

A number of people expressed interest in this work, and I was invited to visit various people in different departments around the USA – invitations that in some cases I was only able to take up between twelve and forty years later!

In a symposium on the teaching of gross anatomy, I was moved to offer the criticism that the speakers seemed to be preoccupied with economic issues, rather than academic and didactic ones. Both Bill Greulich of Stanford and William Gardner of Yale countered my strictures when I lunched with them (see Chapters 15 and 21 respectively). This, they said, must be seen in the light of the country's economy: Young anatomists (like other lecturers) must have enough money. Even if young idealists emerged, their wives would soon insist on washing machines, 21-inch TV screens, and the like. 'It is,' they said, 'regrettably inevitable.' As one of the idealists of this world, this saddened me deeply. They seemed helplessly caught. I recalled the story told to me by Burton Baker: he was a member of a team sent to a Liberal Arts College at Grand Rapids, to tell the students about anatomy and its attractions. When he asked, 'Are there any questions?', the first one was: 'What could I make once I have my doctorate?' The second question was: 'What is the most I could make?' At that point Baker stopped in disgust and gave them a sermon.

I cannot end the story of my sojourn in Milwaukee without mentioning what heralded the arrival of the anatomists in the city. As we approached, we saw several huge balloons hovering over Milwaukee. The appended sign, which could be read from kilometres away, was THE DAY THE BODY-SNATCHERS CAME TO TOWN! It referred to a film of that title that was showing in the city, but the double entendre appealed greatly to my sense of humour.

I HEADED SOUTH for the next phase in my travels. Apart from my time spent at the Birmingham Medical School in Alabama, I was taken to Tuscaloosa, the scene of the main University of Alabama campus. The town was small, with about 50 000 inhabitants, neat-looking, hot – it reminded me of the small university town of Potchefstroom in South Africa. The University of Alabama was celebrating its 125th anniversary. Its protestations of service to the community, over the entrance gateway – 'For peace, equality and democracy' or a similar catalogue of 'do-good' purposes – rang a little hollow, for this had been the scene of a race riot some six weeks previously, when a court order instructed the university to admit Autherine Lucy, a black lady, as a student. The resistance to her presence on the campus, the flare-up of student violent opposition, probably egged on by townspeople, the refusal of the university authorities to enrol Ms Lucy, precipitating a second court order that she be admitted – all this marked a turning-point in the civil rights movement. I gazed in horror and amazement, exquisitely mindful of the apartheid government's threats against the presence of students of colour at Wits and against the Open Universities themselves; mindful, too, of the struggle we had been waging for some eight years against this racist sabre-rattling. I was not to know that the long-dreaded and -resisted legislation was looming ever closer – it was but three years away. So while American society and the civil rights movement were hovering on the brink of a jubilant breakthrough to a non-racial and non-discriminatory accomplishment, in South Africa, we were poised on the verge of another kind of leap, an inglorious one, which would set our universities and the country back for 30 years or more. Two mighty processes met at that moment, one moving forward hopefully, one retrogressing ignominiously.

From Birmingham, I sallied forth by two planes and a bus until I reached Cincinnati, Ohio, at 11.00 p.m. to be met by my hosts, George Barbour, and his wife Dorothy. He was dean of the College of Arts and Sciences and was also an adviser of the Wenner-Gren Foundation and Leighton Wilkie's Do-All Foundation.

Over the next few days, as I listened to Barbour's wealth of stories and comments about various colleagues and associates, I was confronted once again with the quibbling and backbiting that seemed to be characteristic of the big international centres that I had visited. I could not help thinking, as I had thought before (and have often since), what an advantage it was to be working in Africa, relatively far from the pettifoggery and the wrangling. For the whole of my life in science, I have tried to eschew such squabbles and if any of my staff or research students have come to me with queries of this kind, I have quoted that good formula that 'Jean' wrote in my autograph book when I was just eleven years old (see Chapter 1):

> Don't look for the flaws as you go through life,
> And even if you find them,
> 'Twere wise and kind to be somewhat blind,
> And look for the virtues behind them.

On a visit to the Cincinnati Anatomy Department, I spent time with the head, Professor Crafts, formerly of Maine. Amongst other topics, we discussed the nationwide shortage of cadavers, a phenomenon that was threatening to halt the essential practice of dissection.

There was a move to exhort people to bequeath their bodies after death, but Crafts thought he detected resistance to this when people ascertained that the cadavers were kept in tanks. As long as this practice persisted, people were not going to be willing to bequeath their bodies to the Anatomy Department.

Consequently, Crafts was experimenting with an alternative method: after embalming, a body was put into a bag of a compressed, non-porous cloth, and clamped all around, thus sealing it. He had had good results.

Tanks were still in use at Wits, but soon after I took over from Professor Dart in 1959, I planned a new approach: after embalming, each body was bagged in a watertight and airtight container. These in turn were placed in stainless steel cabinets, each on its own shelf that could be drawn out on casters, as has come about in modern mortuaries. We have never had any complaints from would-be donors or bequeathers. In 1959, I made an appeal for public-spirited citizens to add the appropriate codicil to their wills. To make it easier for them, we worded a legally appropriate codicil. Within a very short time, we had some 3 000 names on our books, of people who had informed us that they had added the codicil to their wills. Shortages of bodies became a thing of the past. The Anatomy Act was amended and greatly liberalised, in keeping with proposals made by my department. In fact, I don't think it would be too much to claim that it was the only liberal piece of legislation to be passed by the South African parliament during the apartheid regime!

THE FOUR OR five days I spent at Western Reserve University in Cleveland, Ohio, were teeming with interesting ideas and people. The 'Western Reserve Scheme' was on everybody's lips and I learned about it from various people. A basic principle was that teaching modules no longer fell under independent departments, but under inter-departmental committees. The purpose of this was to achieve a greater measure of integration than there had been before. Members of committees attended one another's lectures and criticisms were given. Through the committees and through sitting in on one another's lectures, staff members were becoming healthily critical, not only of one another, but self-critical, too. They were taking tips and becoming better informed on borderline subjects and aspects.

I rejoiced that at last here was a mechanism for teaching the teachers how to teach. At Wits, in a system started by Dart and developed and expanded by myself, we used the attendance of staff members at seminars by third year B.Sc. and B.Sc. Honours students for the same purpose. The critical comments following the delivery were divided into two sections:

those on the presentation and those on the content. It was during the comments on presentation that many points emerged that were of great value to lecturers, although they were directed in the first instance to the student presenter. There were points such as audience contact and eye contact; not allowing your notes to be a barrier between you and the audience; lifting your voice and projecting it to the back row; addressing the congregation and not the chalkboard; repeating, spelling and writing up new and difficult terms; laying out the order of presentation – I used to say: 'Tell them what you are going to tell them; tell them; and end by telling them what you have told them!' (I learned that from Bradley Patten at Ann Arbor.) If your audience is taking notes, moderate your speed of delivery to that of the slowest ship in the convoy. These were some of the tenets that our Wits anatomy staff learned from their attendance at seminars. Later, I went further and, as head of the Department of Anatomy, I sat in on lectures by new, junior and inexperienced staff members and then would have a private discussion with the lecturer in my office. It is undeniable that the competence and enthusiasm of a lecturer has a direct bearing on the students' experience of a subject – and the lecturer needs to take this responsibility seriously.

20 | Hot and humid days in Washington

In Washington, DC I was met at the airport by Professor W. Montague Cobb, the head of the Department of Anatomy at Howard University. 'Monty', as everyone called him, was 51, an intellectual, something of an old-time morphologist, a keen believer in and worker for civil rights, an officer in the Medical Committee of the National Association for the Advancement of the Colored People (of which he was for a number of years president). He was an avid reader and writer, and a violinist.

Monty was an indefatigable host. After settling my things and me at the hotel, and personally calling up the proprietor (it was 12.15 a.m.) to ensure my comfort, he insisted on driving me around the sights of the city. Then he said, 'You must come up and meet the wife' (it was 1.00 a.m. by then). His wife was still up, marking school-papers, and his mother-in-law had waited up for my arrival. We drank beer and munched on breadrolls and chunks of cheese for some hours.

Mrs Cobb was a teacher at a predominantly Negro (African-American) school. She told me something about the progress of desegregation in education, which seemed to be more advanced in Washington than in many other places.

I was very interested in matters educational and had played an active part in the Education League of South Africa, of which I had been chairperson for several years. The League was dedicated to fighting against the policy of 'Christian-National Education' (CNE) that was being propagated by the apartheid government and its minions. CNE stood for segregation at schools and universities, not only between black and white learners, but also between English-speaking white pupils and Afrikaans-speaking white pupils. In all matters educational the CNE policy was obscurantist with mediaeval overtones. Despite the title of the movement, the Education League asserted that what the policy stood for was neither Christian nor

National – nor, for that matter, Education! As I had been actively involved in several movements to expose and oppose this threatened educational system for close on a decade, I was most interested to learn about the American experience in desegregation and it was particularly illuminating to discuss the issue with Mrs Cobb who was herself coloured in the American sense.

While I was chatting with her, Monty disappeared into another room from which we heard exotic, late-night strains emerging. 'He always practises on his violin at this time,' said Mrs Cobb. It was 3.00 a.m. by the time Monty dropped me back at my hotel.

MY ADDRESS TO the medical students and staff at Howard University Medical School found a readier and more responsive audience than almost anywhere else. How hilarious they found my jokes to be, how uproarious was their response. Monty, in introducing me and especially in thanking me at the end, stirred the pot deliciously, especially when in true anatomical and anthropological style, he named 'Tobias's angle of dangle' – an allusion I am not going to explain, even though I am too old to blush!

A few weeks later, on 24 May, I returned to Washington, DC from Boston. I was suffering from a raging attack of sinusitis and was so muzzy, I was barely conscious of flying or of the three stops en route. When I landed at Washington, Monty was again there to meet me. He took me to his new home and left me preparing my talk for that evening and trying to clear out my sinuses. The temperature and humidity were both high. A dinner in Baldwin Hall at Howard University preceded my address. Those present were almost exclusively African-Americans, who were part of the medical and other faculties and other physicians. They were members of the Medico-Chirurgical Society of the District of Columbia (founded in 1884). I was to deliver the 12th Annual Charles Sumner Lecture. This was my first eponymous lecture, though by no means my last. (The list now stands at some 60 eponymous lectures, which I have delivered at home and abroad.)

The Charles Sumner lectureship had been set up to commemorate Senator Charles Sumner, an American senator who, said Monty, 'made valiant efforts over a period of years following the Civil War to have racial bars against Negro medical practitioners removed by statutory means in the District of Columbia'.

I was the first foreigner to speak as a Charles Sumner lecturer. My topic was 'Hunters and medicine men of the Kalahari'. Apart from some difficulties with the slide projector, it went off very well indeed. I didn't start speaking until 10.30 p.m., and the enthusiastic audience kept me on the go until 12.30 a.m. This is the latest hour to which I have ever continued speaking and fielding questions.

Afterwards, I went with Monty to the medical school, where his staff had been setting a 'spot test' or 'cadaver walk' as he liked to call it. The little flags and labels were on the students' own cadavers and every table had a bone or bones on it. Then, in the wee hours, we went for a drive along the Potomac River, through one of Washington's beautiful parks, willows hanging over the water on which sparkled a thousand reflections of the full moon. At intervals of around 25 metres were parked cars containing embracing couples. It was nearly 2.00 a.m. when we went in to a seafood emporium for refreshments – and then home to bed. After less than three hours' sleep, I was astir, making for the airport and a sleepy return flight to Boston.

From my first visit to Washington, DC, a close bond was struck between Monty and me, which was to last for many years. In 1957–58 when Monty was president of the American Association of Physical Anthropologists, the association awarded the 1957 Viking Fund Medal in Physical Anthropology to Raymond Dart. The medal was presented to Dart by Cobb in March 1958 and I recall Dart being very touched by the thought that he, a white South African, received the medal at the hands of Monty Cobb, a black North American.

Then, in 1977 Monty was awarded an honorary doctorate of science by Wits University. He was the first African-American to receive an honorary degree from Wits and the university gave me the honour of presenting him for the honorary degree and reading the citation I had prepared.

Monty again came to South Africa in 1984 to attend the International Symposium on the 60th anniversary of the discovery of the Taung skull at Wits, Johannesburg and Mmabatho. He was one of three octogenarians who came from the USA for that memorable congress (see Chapter 17): the others were Mildred Trotter and T. Dale Stewart. Sadly, that was the last occasion on which I saw Monty who died in 1990.

IT WAS ALSO on my first visit to Washington that I met T. Dale Stewart, the curator of physical anthropology at the Smithsonian Institution. Stewart took me to lunch with his colleagues and spent much time discussing problems with me and showing me round the museum. He also arranged for me to give an address to members of the museum staff, about 40 or 50. Finally, he took me to spend the weekend in his delightful country home in Virginia, not far from the Potomac River. Everything was resplendent in springtime bloom: there were masses of azaleas, dogwood (a white-blossomed tree), 'Jack in the Pulpit', Virginia creeper, poison ivy – and the woods were teeming with plant, bird and insect life.

The air was so heavy with pollen from the encircling trees and shrubbery that I had very bad sinus trouble the whole time I was there. This constitutional weakness of mine had already been touched off by the pollens, heat and humidity of Birmingham, Cleveland and Cincinnati. The psychological effects then – and indeed throughout my life – were marked. My sinus problems put a definite damper on my activities, clamped down on my clarity of thought (such as it was), marred my speech and questioning, and made me disgruntled. Suddenly, after four months there, and with six weeks to go, I had become rather acutely fed up with polliniferous America, and was sick and tired of travelling. It is interesting how marked this somatopsychic effect can be. I would find myself speaking almost automatically, not registering any very concrete impressions and delivering well-received addresses almost as though by second nature.

STEWART HAD BEEN a protégé of the formidable Ales Hrdlicka, his predecessor at the National Museum. Of Czech origin, Hrdlicka had come over to America around the beginning of the twentieth century, after training under Léonce-Pierre Manouvrier, one of the critical figures in French physical anthropology in the nineteenth and early twentieth centuries. A small man, Hrdlicka had a rather fiery temperament – in Stewart's words, he was a real prima donna. Whatever he went into, he was driven to dominate and ultimately lead. He tended to set himself up as a sort of arbiter at meetings and often delivered himself of scathing criticisms of papers on these occasions.

Hrdlicka started the *American Journal of Physical Anthropology* and, until the Wistar Institute took it over financially, he often financed it out of his own pocket. Hooton, with whom he always remained on friendly terms, had been one of the first subscribers to the *Journal* and substantial help had been received in the early days from Mrs Hooton, who had apparently been a wealthy woman in her own right.

Hrdlicka had been largely responsible for the breakaway of American physical anthropologists from the anatomists. According to Stewart, this was based not so much on any logical reasoning, as on Hrdlicka's personal antipathy towards, especially, T. Wingate Todd.

Hrdlicka may have been a harmful influence in some ways: he deliberately and systematically poured cold water on every attempt to establish any sort of antiquity for man in the New World. This did a great deal to dampen the enthusiasm of younger scientists.

Unlike Hooton, who did very little fieldwork, Hrdlicka did a great deal, especially in Alaska. His excavations there yielded many skeletal remains, which now constituted one of the biggest single sections of the Smithsonian collection of 18 000 human skulls. Hrdlicka went back into the field every year or two until late in his life, which ended in 1943.

Dale Stewart told me that much of his own current research was of a routine or opportunistic character, as material came to hand. Collections of bones from ossuaries were constantly being sent in and required identification and description. Then he told me how he had been over to Japan, making researches on the bones of hundreds of young Americans killed in Korea. Each body had been buried with a small sealed canister, containing details of name, age, origin, and so on. Afterwards, when the enemy forces had overrun these burial grounds, they had dug up many of the skeletons. They had not been as punctilious about keeping bones and records together, as the scientists would have been. So every skeleton had first to be identified.

The American Army had run the entire project with a minimum of publicity: it was hoped thereby to avoid any unpleasant reactions from the relatives of those who had died on the field of battle. It seemed to me that the ethics of this situation were complicated, but perhaps not as complicated as those involved in the German scientists' experiments on victims of the concentration camps. Nevertheless, I could not help wondering whether the American Army's decision to have the remains in question studied in Japan and Hawaii, and *not* after their removal to mainland America, smacked of duplicity or not.

Stewart's methods were fairly close to the orthodox biometrical approach, but he had much more morphology at his command and more of a feel for biological and pathological processes than, for example, Jack Trevor at Cambridge, who at that time represented the 'purest' extant example of a biometrician known to me.

Stewart had clashed several times with Sherry Washburn of the Department of Anthropology at the University of Chicago. There was an encounter in Chicago that I witnessed at the AAPA meeting in 1956. Stewart, a previous editor, and Washburn, the current editor, of the *American Journal of Physical Anthropology*, disagreed vigorously over Washburn's proposal to change the name of the journal to the *American Journal of Human Evolution*, as well as over Washburn's pride in proclaiming that no articles submitted to the journal had been rejected under his stewardship.

Sherry gave the impression of having cut himself off, very consciously and pointedly, from all – or virtually all – earlier work in his subject. Even on those aspects on which he had worked, he did not know – or professed not to know – the earlier work. It was this denial of past work that had led Henri Vallois to comment that Washburn's 'New Physical Anthropology' might be 'anthropology' but it certainly was not 'new'.

In the same vein, I recall my astonishment at a remark made by Washburn in commenting on one of my seminars to his graduate students: he noted, apparently with surprise, how a 'new' and 'peculiarly South African' approach in physical anthropology was growing up, drawing its sources not from Sir Arthur Keith and England, but from Bolk, Sergi, Frasetto – people I had cited in my seminar. Yet, it did not strike me as at all odd: my philosophy was that, with physical anthropology so young and relatively restricted a field, a scientist should quote not only English-medium but also foreign workers who had made relevant and significant contributions. I should have thought it inescapable to take their views into consideration. Science does not function in a vacuum. Stewart mentioned to me that one of his students had written to Washburn on a subject on which the latter had published: the student had wished to do some research on a similar line and asked Washburn for references to earlier work. Washburn replied that he did not know any. When I think back on my career, my personal assistants will testify how frequently I have received such letters seeking earlier references. In every case, I would not rest until I had unearthed the answers, so much so that I once jokingly described myself as a one-man information bureau on a myriad topics ranging from anatomy to apartheid, growth, sexual dimorphism, somatotypes, ancient hominids from various parts of the world, living peoples of Africa, brains and brain evolution, dermatoglyphs, the suntanning potential of light and dark peoples, sex, race, racism, this, that and the other!

What struck me as alarming about all this was that Washburn was training a great proportion of the future physical anthropologists of America, and that these folk were being given a 'party line', with little or no attempt to acquaint students with the views and works of other authors, especially foreign scholars and earlier contributors. It was this aspect of the 'New Anthropology' that gave me most cause for alarm.

NO VISIT TO Washington by a person with my interests would be complete without a journey to the National Institutes of Health (NIH), Bethesda. This complex was beautifully situated in parkland and comprised a number of Institutes devoted to such varied medical topics as Heart Diseases, Arthritis, Mental Health, and so on. Over 3 000 people worked there. The Clinical Center of the NIH was a great, elongated, fourteen-storey building – a sort of backbone, with appendages going off on both sides. The 'backbone' was the hospital, to which only hand-picked patients were admitted, free of charge, while the appendages were research rooms. There was more space for research than for clinical care.

One section was for the medical arts, which planned and executed demonstrations, exhibits and illustrations for papers. They had some fine techniques and I took some literature and references from them, which I was able to put to good use in the Wits Anatomy Department.

I was also able to do a fair amount of sightseeing while in Washington. I saw the Episcopalian Cathedral, which had been under construction for 50 years and was expected to take as long again to complete. It was a magnificent neo-Gothic edifice with beautiful examples of iron and wood craftsmanship, set in lovely gardens near the Bishop's residence. I attended a promenade concert in one of the garden courts of the Art Gallery amid playing fountains and banked flowers. Dale Stewart and I also attended an excellent Cole Porter stage show, *Can Can*. We drove through Georgetown, an original early settlement, now incorporated into Washington. Many of the old dwellings had been rebuilt in Georgian style and were simple, plain, but with impressive dignity. We saw hundreds of fishermen and fisher-boys on the rocks of the Potomac River, picnicking and catching baskets full of fish. As we crossed the bridge over the river, the fish could be seen rising in the water!

Whilst I was in Washington, we had excessive heat-waves. On two days running the temperature reached 92 °F (33.3 °C), and the humidity was awfully high. I had not felt anything so hot and sticky since my Durban days, and my sinuses continued to be congested and troublesome.

21 More East Coast travels

My long and happy association with the Anthropology Department and Museum of the University of Pennsylvania started in May 1956. It continued for just over 40 years, ultimately seeing me become a visiting professor, an eponymous lecturer and the recipient of a Doctorate of Science *honoris causa*. Way back in 1956, one of the first considerable personalities I visited and befriended was Professor Loren C. Eiseley.

Eiseley was a big man, dark and broad, with sensitive and brooding eyes. He was a gentle, poetic soul, cast in a very different mould from any other physical anthropologist I met in America. His prime interest lay in the historical aspects of anthropology, on which he wrote with a free and mellifluous style. To read Eiseley was to be at once enlightened by the depth and range of his conceptual world and enchanted by the loveliness of his word magic. At the time of my visit he was editing some letters of Charles Darwin that had recently been acquired by the American Philosophical Society of Philadelphia. He was working also on a book for the Darwin Centenary Year (1959). His literary *oeuvre* was extensive and included fragments and essays for *Harper's Magazine* and other publications, usually with a philosophical bias. These essays betrayed a personality alive to the beauty and wonder of nature, youthful in its ability to appreciate a bird's wing in the midst of a great city, exquisitely sensitive and, withal, of a somewhat sad and melancholy outlook.

He inculcated a love of the history of the subject into his students (most notably Jacob Gruber). He was, I believe, fulfilling a valuable function, since, as I have already said, it was all too common in America – and elsewhere – for scientists to become oblivious of their scientific antecedents. Yet it was foolhardy to strike out for new directions in anthropology, without a firm or even acknowledged root in the past. In other words, Eiseley's approach was the very antithesis of Washburn's.

Among others in the Anthropology Department was Dr J. Louis Giddings who was working on Inuit (Eskimos). He had a similar difficulty of definition to ours with the Bushmen or San. He had concluded that Eskimos were best judged on linguistic grounds. Or, if living Eskimos recognised *most* of the artefacts in an assemblage, the culture represented was considered to be Eskimo.

Dr Ward Goodenough, with whom I became most friendly in later years, had worked on New Britain, the largest unit in Melanesia (100 by 485 kilometres). He was labouring assiduously in the quest for synthesis at the interface between biological and cultural anthropology. Years later, when I was heavily involved in studying the evolution of language, I had many fruitful discussions with Goodenough on the number of speech sounds accompanying a recognisable language. When I claimed that spoken language had first appeared about two million years ago, in *Homo habilis*, on the basis of my studies on endocranial casts of very early hominids and on associated cultural accomplishments, there were colleagues who disputed this. It was Ward Goodenough who demolished their primary criticism by pointing out that in a number of Pacific island languages, the people – and their language – gained their linguistic ends with an extraordinarily small number of speech sounds – yet, for all that, they qualified as languages. Goodenough was an active member of the distinguished academy, the American Philosophical Society; the most prominent of whose founders in 1743 had been Thomas Jefferson. It was on Goodenough's nomination that I was elected a Foreign Member of the Society in 1997.

Carleton S. Coon was another of the unforgettable characters in the Anthropology Museum at the University of Pennsylvania. Bluff, hale, jovial, hard-hitting, his character was well epitomised in a photograph taken at Livingstone, Zambia, during the third Pan-African Congress on Prehistory in 1955: it showed a notice that read, 'All Wild Animals are Dangerous' – and Coon leaning on it, scowling! He was busy writing books: his *History of Man* was just out and another to be entitled *Seven Caves* was on the way. In addition, he was off to Japan to look for traces of Palaeolithic man there. He told me that the Japanese had never really searched for fossil ancestors because this would have been in conflict with Shintoism and Emperor-worship – that is, if the remains of ancient people had been found to have existed in Japan before the Emperors. So he was going to have a look around.

On my visits to Japan in the 1990s, I found that Coon's view was essentially corroborated by what I learned from Japanese anthropologists, especially the younger ones who had grown up under the new post-war Japanese constitution. One or two of the older scientists were still resistant to the idea that humans had migrated into Japan a long time ago. This was a fascinating example of the impact on a science of a political system whose tenets were at variance with some of the hypotheses of the science. In this case, the hypotheses were that humans had entered Japan from the East Asian mainland, or via some of the strings of islands connecting Japan to the mainland to the north, the west and the south – earlier than the Holocene and perhaps in the earlier part of the Later Pleistocene or even in late stages of the Middle Pleistocene. The supersession of imperial hegemony in the second half of the twentieth century, opened the way for Japanese scientists to delve into the prehistoric past of Japan – with fruitful results.

At that stage, 1956, Carl Coon had not yet published his controversial book, *The Origin of Races*. Indeed in the very year of my visit, he was able, thanks to an Air Force contract, to make a seven-month trip around the world, collecting the data that were to be enshrined in *The Origin of Races* (1962, 1963). By 1959, it had become clear to him that 'the visible and invisible differences between living races could be explained only in terms of history. Each major race had followed a pathway of its own through the labyrinth of time. Each had been moulded in a different fashion to meet the needs of different environments, and *each had reached its own level on the evolution scale*'.[9] The part of this that I have italicised reflects a pattern of thinking that is encountered throughout the book. To most human biologists, the idea that each race had 'its own level on the evolutionary scale' is not acceptable. Such argument is the very basis of racialistic thinking, or as it came to be called later, racism. In a word, Coon held that *Homo erectus* had evolved into five races of *Homo sapiens*. He claimed that Africans had evolved more slowly than other major races and this generated great and biting controversy. Apart from these racist overtones, other critics took issue with Coon's grasp of evolutionary mechanisms. Always controversial, Coon's standing and reputation were never the same again after that book appeared.

THE WEEKS THAT followed, which I spent in Boston and Cambridge (Massachusetts), slipped vibrantly past. Although I had had some wonderfully stimulating days and weeks at other American centres, Cambridge/Boston was without doubt one of the highlights. Seldom have I encountered in one area such a concentration of talented and brilliant people and so charged an academic atmosphere. Although by this stage I was spent and in need of a rest – after nearly six months of unbroken travelling, conversing and lecturing – my batteries were soon recharged in the New England environs. This recrudescence of strength was to come back time and again on later visits to Cambridge and Boston.

Apart from Harvard, with its main campus in Cambridge and medical school in Boston, there were the Massachusetts Institute of Technology, Boston University, Boston College (Catholic), and Brandeis University (a new, Jewish-sponsored university just outside the city, where, around a big pool, were Hebrew, Catholic and Protestant chapels).

Boston and New England generally showed – and they liked to show and to cultivate – a solidity and a maturity that I did not find elsewhere. The 'Ivy League' Universities – Harvard, Yale, Columbia, Brown, Cornell, Duke, Pennsylvania, William and Mary – were founded in the days of the British colonies and they had definitely acquired a poise, dignity and reliability that were not as evident in other establishments. The way of life, too, was more relaxed, more leisurely, in many ways more 'British' – even down to afternoon tea served in bone china cups from silver teapots. New England and the rest of the country seemed to cherish a mutual scorn – the very phrase 'Ivy League University' is curled off the tongue of a mid-Westerner with barbed contempt, while New Englanders had continued until fairly recently to regard everything west of Boston as 'Wild West!'

A HIGHLIGHT OF my stay in Cambridge, Massachusetts, was my visit to Lawrence and Lorna Marshall. He was a white-haired, robust, elderly, retired businessman, having been in some sort of electronic business. Lorna was a seemingly frail, grey, little woman, though full of life and knowledge and trained in anthropology. Their son John was a senior at college. The daughter Elizabeth was then living in North Carolina, her husband about to complete his military service. In 1950, Lawrence Marshall, not having had

Phillip at fourteen months of age with his sister Valerie Pearl Tobias and their mother, Fanny Tobias (née Rosendorff), 3 January 1927.

At two years old Phillip was scarcely bigger than a cricket bat.

Fanny and Bert Norden on their wedding day, 25 February 1937. Albert Louis Norden had been Phillip's godfather and now became his stepfather.

Rowing during a wolf cub camp at Mazelspoort near Bloemfontein (1938). In the background is Phillip's cousin Walter Posner.

The family in Durban, July 1937. From left: Val and Phillip, Bert Norden, Fanny known as 'F.', and cousin Harry Posner.

St Andrews School, Bloemfontein, in 1938: Form 4A (the equivalent of Standard 6 or Grade 8). The form master Roger C. Bone is in the centre of the second row with Phillip on his right. Phillip's first editorial activities occurred in that year, when he edited the *Form Four Fortnightly*, which was typed by Bone.

The Medical B.Sc. class of 1945 in front of the Wits Old Medical School on Hospital Hill, Johannesburg. Back row, from left: Elizabeth Westphal, Maureen Dale, Margaret Cormack, Sam Axelrod, and Abdul Hak Bismillah, partly obscured behind Phillip. Second row: Samuel W. Hynd, Anthony C. Allison, Priscilla Kincaid-Smith, Dennis T. Glauber and Penelope Stewart. Front row, Bunny Tabatznik and Joseph Stokes Jensen.

Phillip in 1942 as a prefect at Durban High School. The prefect's badge on his lapel is now in the school's museum.

Joseph Gillman, Phillip's most stimulating mentor during his Science and Honours years in the middle to later 1940s and supervisor of his Ph.D. thesis.

Phillip with Sydney Brenner at lunch one day in 1946
at St Margaret's, Johannesburg.

The famous 1945 Medical B.Sc. class expedition to the Makapansgat Limeworks that collected monkey and baboon fossils and effectively catapulted Raymond Dart back into the field of palaeo-anthropology. Phillip was the organiser and leader (2nd from right in front row) and the deputy leader was Joe Jensen (2nd from left in front row).

The Student Assembly of the National Union of South African Students (Nusas) at the University of Cape Town, July 1949. Many famous student leaders and distinguished figures in academia in later years are to be seen, such as Sydney Brenner squatting in front (right), Wolfie Traub (right end of front row), Bernard Harley Kemp (left end of front row), Hans Meidner (right end of back row), and Harold Wolpe (left end of the second row). Phillip, the president, is in the middle of the front row.

Phillip (aged 25) with his father, Joseph Newman Tobias (aged 66), on the occasion of Phillip's obtaining his medical degree at Wits, 6 December 1950.

Phillip with his mother, 'F.' Norden (aged 59), on the same occasion. The two academic hoods he is wearing are those of B.Sc. Honours (with a stripe) and the MB.BCh. (grey, partly folded).

The French Panhard-Capricorn expedition sets out from Germiston for Namibia and Botswana. Phillip with beret and standing next to Francois Balsan, the expedition leader, and with other members of the project that gave Phillip his introduction to the living Bushmen (San) of the Kalahari in 1951.

Arriving in Maputo (then Lourenço Marques) in October 1951, towards the end of the Panhard-Capricorn expedition.

Phillip becomes a full-time lecturer in anatomy under Raymond Dart in 1951.

Phillip's beards waxed and waned with the coming and going of field activities.

Phillip with 'Oubaas', a hypopituitary midget of the Makoko population of Bushmen in the Ghanzi district of Botswana.

With J. Lawrence ('Larry') Angel in Philadelphia in May 1956. Larry seconded Phillip's nomination for membership of the American Association of Physical Anthropologists (AAPA). The photograph, one of very few taken in 1955 and 1956, depicts 'Tobias and the Angel'.

Father Franklin Ewing S.J. holding the neandertaloid fossil 'Egbert' at the entrance to Fordham University, New York City, in June 1956. Egbert had been recovered by Ewing from the Ksâr 'Akil cave in the Lebanon. Sadly Egbert was subsequently lost or mislaid in the process of repatriation to the National Museum in Beirut.

Phillip with Martin Bobrow (at left) and Laurie Geffen (centre) at the Kariba wall in 1957. This first human biological survey was in the Gwembe Valley, among the Tonga people who were to be displaced above the rising waters of the man-made Kariba Lake.

Setting out from the Wits Medical School for the second leg of the Kariba Tonga Survey in 1958. Left to right: Adrian Martin, Phillip, Richard van Hoogstraten and Dawood Mangera.

Desmond Clark (left) and Louis S.B. Leakey on a little bridge near the Congo River, during the Pan-African Congress on Prehistory and Palaeontology in 1959. At this meeting in Kinshasa Leakey announced the discovery (by his wife Mary) of the cranium of *Australopithecus (Zinjanthropus) boisei.*

Phillip with little Zambali at Mwemba on the bank of the Zambezi River, during the human biology survey of the Tonga people who were living below the waterline of the future Kariba Lake.

Raymond Dart and Phillip at the Wits Graduation Garden Party in April 1959. At the beginning of that year, Phillip succeeded Dart as head of the Wits Department of Anatomy. (Photo: *The Star.*)

Phillip with Louis Leakey and 'Zinj' or 'Dear Boy' during the Leakeys' visit to the Wits Anatomy Department soon after Mary's remarkable find in July 1959.

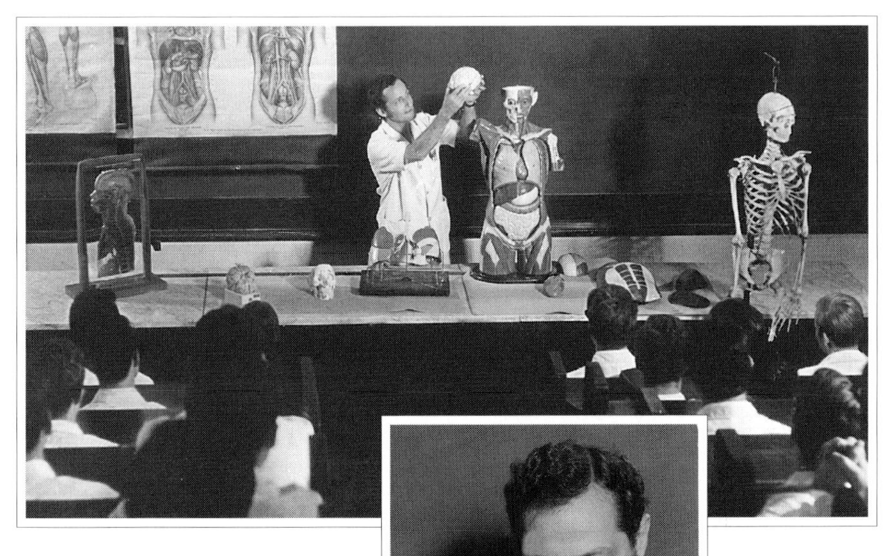

Teaching anatomy to medical and dental students in the Vesalian Lecture Theatre of the Old Medical School. The old-style props are very evident, and Phillip learnt this technique from Dart.

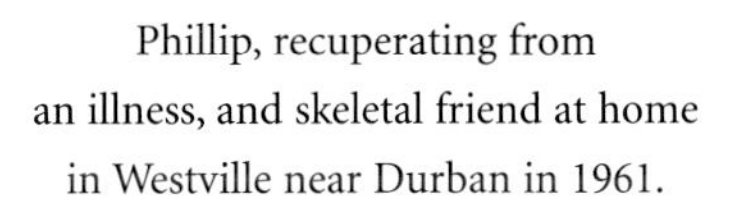

Phillip, recuperating from an illness, and skeletal friend at home in Westville near Durban in 1961.

An Afro-Asian conference with a difference! From left to right: Jack C. Trevor, Phillip and Ralph von Koenigswald in the Duckworth Laboratory, Cambridge, in June 1964, with original fossil hominids from Tanzania and Java. (Photo: Len Morely.)

Ralph von Koenigswald in his laboratory at Utrecht in August 1964 – lovingly holding a specimen of 'Java Man'.

Lord Zuckerman comes to visit *Australopithecus* – at long last – in 1975! In the fossil laboratory with (left to right) Ron Clarke, Lord Zuckerman, Alun Hughes and Phillip. Here, Phillip shows him how the brain size of *Homo habilis* was determined. Zuckerman remained unconvinced about the hominid status of *Australopithecus*.

George Nurse (at left) and Trefor Jenkins whose important contributions on health and the San hunter-gatherers are celebrated by Phillip in the text.

W. Montagu Cobb comes from Howard University, Washington, DC, to Wits to receive an honorary doctorate – the first African-American to be so honoured. Phillip was his promoter and read the citation.

Theodosius Dobzhansky, 'the Darwin of the twentieth century', who was the external examiner of Phillip's Ph.D. thesis and contributed the Foreword to his first book. This picture was taken by Bob Brain in Doby's laboratory at the University of California, Davis. In the background are two icons of Doby's career – *Drosophila* (the fruit-fly) and Charles Darwin.

Sir Wilfrid Edward LeGros Clark, FRS (centre), with the vice-chancellor and principal of Wits University, I.D. MacCrone (right) and Phillip. The photograph was taken on the occasion of Sir Wilfrid's visit to Johannesburg to deliver the 2nd Raymond Dart Lecture and to receive the degree of Doctor of Science honoris causa from Wits University in March 1965. Sir Wilfrid was an external examiner of the materials submitted by Phillip for his D.Sc. degree in 1967.

Robert Broom (left) and Peter van Riet Lowe examine a tooth of *Equus capensis* in the Cave of Hearths, Makapansgat Valley, in 1937. Eight years later Broom took Phillip and fellow students to Sterkfontein for the first time, while Van Riet Lowe stimulated Phillip and his fellow students to mount their famous winter expedition to the Makapansgat Limeworks and caves. (Photo first published by the Royal Society of South Africa, 1948.)

much chance to get to know his son well, decided to do a trip with him. To make it useful as well, he got into touch with the Peabody Museum at Harvard. They suggested a trip to Bushmanland. So, in that year, he and John made their first trip to Africa, mounting a reconnaissance of the terrain and of the Bushman territories. They decided on a suitable area near Gautscha Pan in South West Africa (now Namibia). They returned in 1951 and then, with a large team, came back for their third and greatest expedition in 1952–53. Under the auspices of the Peabody Museum of Harvard University, they lived for fourteen months continuously among the same group of !Kung Bushmen. This duration was a record for such an expedition. In that time they acquired a profound insight into the !Kung way of life around the cycle of the seasons. There were to be five more expeditions, all of which were sponsored by the Peabody Museum, the Smithsonian Institution and the Transvaal Museum in Pretoria. The fourth had taken place in 1955; the fifth followed later in 1956, while the last three occurred in 1957–58, 1959 and 1961. These eight expeditions were parallel endeavours to the Panhard-Capricorn Expedition of 1951 (see Chapter 10) and the fifteen years' series of expeditions mounted by the Kalahari Research Committee of Wits University (1956–71).

On their first four expeditions, the Marshalls took over 200 000 feet of coloured film. They came into contact with or got to know 600 Bushmen.

Lorna Marshall was most interested in folklore and was writing up her records. One article had already appeared. She begged me to include records of folklore in my expeditions, whatever other research objectives I might have been planning to pursue. Many of the old myths, she feared, were already being forgotten.

This extraordinary family, as their retirement project, added greatly to our knowledge of the Bushmen. It was a privilege to have spent time with them in southern Africa and in New England, just about the time when our own researches in Botswana were dispelling many of the still surviving myths about the Bushmen and, in 1956, in the very year when plans were started to set up a Kalahari Research Committee at Wits University.

I AM SURE MY friends at Yale will find it unforgivable that I spent only one day there towards the end of my 1956 visit. Invitations in New York City and at Cold Spring Harbor, toothache and inflamed sinuses, and the imminence of my voyage home, all conspired to prevent my doing justice to New Haven. In fact, I had breakfast in Cambridge, lunch at Yale and supper on the train to New York on 1 June 1956. What happened to my luggage, which by now was rather heavy, I don't remember.

I spent the day with Dr William Gardner, head of the Department of Anatomy at Yale. I had met him previously at the Anatomical Congress in Milwaukee when he lunched with Bill Greulich and myself (see Chapter 19). In New Haven, Gardner and I spoke about something that had worried me since my arrival in America: the glibness of students. Perhaps, Gardner thought, this might result from high competition, but he feared that the danger was that usage of words was becoming substituted for understanding. This was an all-too-common occurrence. Looking back on his own earlier career, he freely admitted that he had been guilty of it. I wondered if this was perhaps the ugliest manifestation of lifemanship imaginable – for Potter and his gamemanship were becoming popular reading in some circles. 'Our Department,' he said, 'is in competition with other departments, but not with the student as the brunt.'

A feature of the Yale department I had not encountered elsewhere was that the dissection halls were open from 7.00 a.m. to 10.45 p.m. The students themselves were responsible after hours; the last ones checked out and closed up. They were treated as men and women and handled themselves accordingly.

I was not surprised but happy to learn that there was an elective course in terminology and history of anatomy – not surprised because of Yale's world-famous Library of the History of Medicine. The nucleus of this library came from three private collections: those of Harvey Cushing, the eminent American neurosurgeon of Johns Hopkins, Harvard and Yale; John Fulton, the former professor of physiology at Yale and at the time of my visit the professor of the history of medicine; and Henry Sigerist, the great historian of medicine, at first of the Medical History Institute at Leipzig and then Director of the Institute of the History of Medicine of Johns Hopkins.

The resulting combined library gave Yale the nucleus of the world's greatest collection of works on the history of medicine. A peep into the library was all that time allowed me, but the highlight was a glimpse of the locked and barred Room of Vesaliana. The momentous work of the anatomist Vesalius was called *De Humani Corporis Fabrica* (1543). Originals of this work are as scarce as hen's teeth. I knew of only two copies in South Africa, one a first edition and the other a second edition, and both were in the private collection of the surgeon-bibliophile, I. ('Oscar') Norwich. Although my personal library is moderate, all I could afford was a facsimile of the first edition. Yet what I peeped into at Yale was a roomful of Vesaliana!

I had always been most interested in the history of medicine (as well as in the history of biology, evolution, genetics and palaeo-anthropology). I often regaled my undergraduate and graduate students with selected topics in these fields. The annual oration at the beginning of the dissection of the body at Wits was always devoted to aspects of the history and philosophy of anatomy and of dissection. One of the three founding fathers of the Yale History of Medicine Library, Henry Sigerist, received an honorary doctorate from Wits University in 1939. He apparently also lectured at our medical school. So greatly were our students enthused by his talks that they set up a History of Medicine Society, supported by half a dozen faculty members who were more than superficially interested in the subject. With the barren days of the Second World War the Society lapsed, but encouraged by myself was re-founded in the 1950s.

At Wits, Dr Cyril Adler, a specialist in physical medicine, and his wife Esther, were pioneers in bringing into being the Adler Museum of the History of Medicine. Cyril and I were deeply involved in encouraging interest by the medical and allied medical students. Most recently, under the deanship of Max Price, the Adler Museum was moved, lock, stock and barrel, to premises opening off the main entrance foyer of the new Wits Medical School. Thus, the Museum has since 2002 been physically integrated into the Wits Faculty of Health Sciences. My own small efforts to advance the Adler Museum since its first inception received encouragement and inspiration from the examples of Cushing, Fulton and Sigerist and of the Yale History of Medicine Library. This alone was a rich reward from my one-day visit to Yale.

I have always been a bibliophile – within the limitations of a poor family background and a South African academic's salary. But, as I often told my research students, buy books while you are young, even if you have to go without a meal. Today a major problem is to find sufficient space for the approximately 25 000 books in my private collection (some 10 000 at Wits Medical School and 15 000 in my apartment). When I lived in Hillbrow, near the old medical school, I ultimately had to buy the flat next door to house part of my collection. The pamphlets, reprints and booklets number another 50 000 to 75 000 items.

The little history of the Yale Library's benefactors has moved me to make a provision in my will for my own collection of works on the history and philosophy of medicine, and some other sections of my library, to be bequeathed to the Wits University and Adler Museum Libraries.

22 | New York City and Cold Spring Harbor

The 'Big Apple' burst upon me in June 1956 with all the impact of the apple tree in the garden of Trinity College, Cambridge, shedding its fruit – according to the legend – on the head of Isaac Newton! Here is an excerpt from a letter I wrote on 17 June to Hertha Erikson (later Hertha de Villiers) who had been appointed as a lecturer in the Wits Anatomy Department that year:

> New York is sweltering hot ... [T]he heat wave ... with humidity to match ... makes my strenuous programme for these last ten days very hard to carry out! Cornell Medical School, Columbia, Sheldon's lab, Fordham University, Bronx Zoo, American Museum of Natural History (where I am lecturing tomorrow) – this is my official list of engagements, not to speak of sight-seeing, shopping, packing. This city is fabulous: my hotel is on Broadway ... from my window I see the crowds in the street below trying to keep cool, and just being plain, natural, bad-tempered but lovable New Yorkers (like the old lady who passed me in the street as I sneezed the other morning; without looking round, 'Gesundheit' she said!); or the dope smuggler who approached me on the corner near Times Square – spreading his 'powdered happiness' (with apologies to Tom Lehrer); or the crazy, mixed-up kids with crazy, mixed-up ids; or the talkative taxi-driver, who grumbles, 'Look at 'em, dashing along, getting, getting, spending, rushing – but d'you think any of them know where they're going? Life's jest one crazy rush!' – while he himself, apparently oblivious, careered through the traffic, rushing, while I clutched my hat and caught at my breath! Or the cigar-smoking, straw-hatted, brow-mopping oldster, who stands in Central Park saying loudly to nobody in particular, 'Heart attacks! Diseased guts! – of course he shouldn't stand ... asking for trouble.' Or the gorgeous, painted, pancaked blondes

> (assisted by 'creative colouring') with dress necklines down to their waists (coming from Africa I felt at home) and heavily-blued lids, hiding those tell-tale shadows of the world's most fabulous night-life; or the old Yiddishe mommas in the Bronx talking to themselves in Russian, Yiddish, German, or just Bronxian . . . while battling along the crowded pavements (oops, sorry, sidewalks) with parcels, perspiration and pants (i.e. their breath); and the hot gospellers of Harlem, rockin' 'n rollin', to the cult of one of their prophets like Daddie Grace ('I believe in G-d the father, the Son and Daddie Grace' – please excuse the sacrilege) . . . and the little girl leaning over the deep freeze and saying exultantly to her friend, 'Mummy's going to let me melt dinner tonight' (or tonite). I am closer to needing that strait-jacket than ever before; but can't bear the thought of wearing one of these new-fangled, American-cut models . . . they're uncomfortable (on my build at least) and anyway they show everything!

I recall that my descriptions were inspired by a midnight temperature of 85 °F, the blare of fire-engine sirens, and the occasional staccato burst of background revolver fire. This letter neatly encapsulates my developing love for New York City. Prior to my last days there, while the city had excited and fascinated me, I still could not claim to have developed a love for it – such as that I felt for London and Paris (a sentiment that, in these two places alone, seemed to transcend the limits of the personalities and human associations at each place).

It took the revelations of my explorations in Greenwich Village, my experience of sunrise in Central Park, the ball-game where I encountered for the first time canned beer and crackerjack, the Basin Street Night Club where I heard and saw live performances by Louis Armstrong ('Satchmo') and Johnny Ray screaming into the microphone (I realised he was hard of hearing), the cleaning out of my sinuses at Dr Grace's Clinic in Brooklyn, the pleasant and urbane welcomes at Fordham, Columbia, Cornell Medical School, the marvels of the American Museum of Natural History, Museum of the American Indian and an electrifying performance of *My Fair Lady* on its first run – it took all these and numerous other sights and happenings to make me appreciate some of the many-sided facets of New York and, with seeing, to awaken my deep affection for the city.

MY ONLY ABSENCE from New York that June was the eight days I spent at the Cold Spring Harbor Symposium devoted to 'The Gene, Structure and Function'. Cold Spring Harbor lies on Long Island Sound. It is a charming, sheltered cove, clustered around by trees and buildings, including the Genetics Laboratories of the Carnegie Corporation and the Center of the Long Island Biological Association.

Over 200 people registered for the 1956 symposium. Wits University was represented by our professor of zoology, the embryologist Boris Balinsky, and myself. Among many distinguished participants was T. Caspersson of the Karolinska Institutet, Stockholm. I had long followed his work on nucleic acid cycles, deoxyribose nucleotides, heterochromatin and euchromatin. In my Ph.D. thesis of 1953 and my 1956 book based on it, I had attempted to wed the results of his studies with my own jejune ideas on my cytogenetical observations.

Another interesting associate, who later visited Wits for our Human Genetics Conference in 1962, was Dr J. Wallace ('Wally') Boyes. He held the chair of genetics at McGill University, Montreal. His main work of late had been in the realm of insect cyto-taxonomy and cyto-phylogeny, and, as I had been doing similar studies on mammalian chromosomes, we exchanged ideas on techniques and results in our related fields.

I gave a well-received talk with my slides on the San or Bushmen. Afterwards a big crowd of us repaired to the Mariners' Tavern, one of several pubs in the village of Cold Spring Harbor, where we had drinks and supper. I was intrigued by the Americans' way of eating oysters – and remarked that I would have liked to have taken a photo of them thus engaged. This occasioned much merriment, as we toyed with the idea of a gathering of Bushmen viewing a selection of slides illustrating the Americans' way of life and eating habits.

Forty-four years elapsed before I next set foot in Cold Spring Harbor. I had been invited to be Distinguished Keynote Speaker at a Cold Spring Harbor Millennium Symposium on Human Origins and Disease in autumn 2000. In contrast with 1956, when Boris Balinsky and I had been the only South Africans present, in 2000 there were several other South Africans, including Trefor Jenkins, Himla Soodyall, Tony Lane and Tony Traill, all of them, as it happened, from Wits.

The year 2000 had not been the kindest of years for me and I had kept the medical and surgical professions rather busy. I had the opportunity to study the workings of two different hospitals, for just over 40 days in all. I remembered anew what wonderfully compassionate, understanding and helpful people doctors are, and the gentle but firm touch of those dedicated souls, the nurses. It was a kind of fun learning to be a biped all over again and graduating, as though in an evolutionary sequence, from six legs to four legs to three legs – and finally to two! Although there were still one or two things that remained to be patched up by year's end, I felt I was able to face up to the challenge of Cold Spring Harbor by late October, but it was especially fascinating to observe that the physical side healed more rapidly than did the spirit. It took frustratingly long for me to have restored my ability to think, speak and write in a positive and creative manner – and I was very conscious of this shortfall when I lectured at Cold Spring Harbor in 2000. By my own standards I was quite simply not at my best, although the audience certainly responded enthusiastically.

IN JUNE 1956, with my dual background in genetics and anthropology (or triune if anatomy be included), I eagerly looked forward to my visit to the Institute for the Study of Human Variation, which was built upon the first two disciplines. I hoped that I would learn more about the synthetic approach to human evolution, so that I could develop a similar philosophy and programme at Wits.

The Institute, about which Dale Stewart had spoken to me, was based in Columbia University, New York City. When it was founded it began to study human evolution. There were several main approaches. Firstly, researchers worked on physical anthropological features, including bodily constitution. The second string to their bow was physiological anthropology, especially on urine and saliva. Needing first to calibrate their methods, they developed paper chromatography of urinary substances; then they applied these methods to monozygotic (identical) and dizygotic (fraternal) twins, families and populations. Thirdly, they worked on gene frequencies of populations. The analysis of normal human variation was a main objective. Twins were being studied from several points of view. An attempt was being made to relate gene frequencies to biology in general.

Against this background of my visit, I was overjoyed when the model set by the Columbia University scholars was followed up for southern Africa by Trefor Jenkins and George T. Nurse. In the course of their research, Jenkins and Nurse worked together on two major monographs. The first, *Health and the Hunter-Gatherer: Biomedical Studies on the Hunting and Gathering Populations of Southern Africa*, was published in 1977. The second, larger book, *The Peoples of Southern Africa and their Affinities*, by Nurse, J.S. Weiner and Jenkins, appeared in 1985. These two works did what I had dreamed of in New York City in 1956, but infinitely better, more elegantly, studiously and comprehensively than I could ever have done. Yet, Nurse and Jenkins most graciously prologued *Health and the Hunter-Gatherer* with an 'Epistle Dedicatory to Prof. P.V. Tobias'. As I have never, on any other occasion, received an Epistle Dedicatory, and as it is not long, I feel sure that they – and you – will not raise any objections to my quoting their letter:

> Dear Phillip,
> It would not be too much to claim that these pages are as much the outcome of nearly three decades of your work as they are of any process of compilation by us. Without the guiding hand of TOBIAS, the biological study of the hunter-gatherers of Southern Africa would have been a much more diffuse and formless thing; but you have laboured to bring order and strict observation into a field previously much bedevilled by mythology, and in the process have been able to contribute not only to a geographically limited subject but to the entire study of man. We hope that you will consent to do us the honour of allowing us to dedicate to you this intimation that although much has already been done in this area there remains, as you often remind us, still very much that needs to be done.
> We remain,
> Your friends and admirers,
> GEORGE T. NURSE AND TREFOR JENKINS

Thank you, dear Trefor and George; your words touched my heart and have remained with me all these years.

ALREADY IN 1956, I had been struck by the number of Catholics and indeed of Catholic clergy who had been steeped in the study of fossil hominids and who had made significant discoveries bearing on human ancestry. There was a scholar such as Boucher de Perthes (1788–1868) whose discoveries had played a seminal part in the founding of European archaeology and palaeoanthropology in the nineteenth century. The Abbé Henri Breuil (1877–1961) was the doyen of French archaeologists in the first half of the twentieth century and was best known for his contributions to the analysis of prehistoric rock art. He worked also in China, South Africa, Namibia and Zimbabwe. As a young student I met the Abbé at the University of the Witwatersrand during his years there and he was one of my advisers when I organised the first all-student expedition to the Makapansgat Limeworks deposits in 1945 (see Chapter 6).

My deep interest in the interrelationships between science and religion led me to the life and writings of Teilhard de Chardin whom I had taken from Johannesburg to Makapansgat, 300 kilometres to the north, and whom a student and I had helped to clamber into the maze of caves making up the Makapansgat Limeworks.

Another significant palaeontologist was Father Edouard Boné of the Catholic University of Louvain, Belgium. An authority on the evolution of the horse family, he visited South Africa several times and worked briefly with Dart on post-cranial bones from Makapansgat, which he and Dart considered to belong to early hominids.

In June 1956 I paid a visit to Father Franklin Ewing S.J. at Fordham University in The Bronx (see Chapter 32). He had excavated in the cave deposit of Ksâr 'Akil near Beirut in Lebanon in 1938. With his colleague J.G. Doherty, he had recovered human remains, said to have been 'neandertaloid' in character. Our meeting gave me a golden opportunity to broach the subject of the interrelationship between science and religion. 'Will you explain the Roman Catholic Church's attitude towards evolution?' I asked him. He gave me an offprint of an article he had published on this subject, following the promulgation of the papal encyclical letter *Humani Generis* issued by Pope Pius XII a few years earlier (1950). The encyclical *Humani Generis* considered the doctrine of 'evolutionism' a serious hypothesis, worthy of investigation and in-depth study equal to that of 'the opposing hypothesis'. It prescribed the attitude of the Catholic Church for nearly half a century.

Then, along with ten other scholars and myself, Boné was a participant in May 1982 in a working group on *Recent Advances in the Evolution of Primates*. This met in the Casina Pio IV in the Vatican. It was organised by the Pontifical Academy of Sciences out of a desire to reconcile data obtained by different approaches. My contribution dealt with the evolution of the brain and spoken language. This topic was to become the basis of a subsequent meeting of the Pontifical Academy, a study week on *The Principles of Design and Operation of the Brain* that was organised in October 1988.

There is little doubt in my mind that the devotion by the Pontifical Academy of these two meetings to the evolution of the primates, including humans, contributed appreciably to the subsequent revision of the Church's earlier defined attitude towards evolution.

Thus, in 1996, Pope John Paul II stated, in his address *Truth cannot contradict truth*:

> Today, almost half-a-century after the publication of the encyclical [*Humani Generis*, 1950], new knowledge has led to the recognition of the theory of evolution as more than an hypothesis. It is indeed remarkable that this theory has been progressively accepted by researchers, following a series of discoveries in various fields of knowledge. The convergence, neither sought nor fabricated, of the results of work that was conducted independently is in itself a significant argument in favour of this theory.

The Pope's statement of 1996 broke new ground by acknowledging that the theory of the physical evolution of humans and other species through natural selection and hereditary adaptation appeared to be valid. The papal message to the Pontifical Academy thus went beyond what had been set forth in *Humani Generis* 46 years earlier, and I am happy to trace my first enlightenment about the Catholic Church's attitude towards evolution to that visit to Father Ewing in June 1956.

MY 80 000 KILOMETRES of travelling since I had left Johannesburg on 30 November 1954 were rapidly drawing to a close. This absence of twenty months from the Wits Anatomy Department and from South Africa has been the longest single furlough from my native country in my entire life. I have served Wits since I was first appointed a demonstrator in 1945 until my formal retirement at the end of 1993; but as an emeritus professor, honorary professorial research fellow and director of a research unit until this day, I have added another eleven years of service to date (and am still counting).

Those twenty months were a turning-point in my career. My life's work had shown enormous progress in the year at Cambridge and my horizons had expanded ineffably. My six months in the USA packed in an incredible number of impressions, visits, contacts and friendships, many of which were to remain with me for the rest of my life. I learned an amazing amount about the very different academic scene in the USA, the teaching methods and course structuring, especially in anatomy, and the interlocking of the various branches of anthropology. I sat at the feet of, and engaged in fruitful conversations with, some of the greatest minds in these cognate disciplines in the USA. At no time in my life has so much been crowded into a mere six months. I visited at least 30 institutions and departments, delivered scores of lectures, seminars, papers – and then, as throughout my life, strove to make no two lectures the same. I cannot begin to tally the lunches and dinner parties, the sightseeing and the flooding of my sensorium with an ever-mutating kaleidoscope of wonders, sounds and thaumaturgy of the mighty America, which, at that time, must surely have seemed near its apogee.

It was circumstances that allowed twelve years to elapse before my next visit to the USA in 1968. By then my love of sailing over the ocean waves had given place to the vibrant tolerance of powered flight. The world had shrunk inestimably.

23 The Tonga people and the Kariba Dam

It was to Desmond Clark that I owed my entrée in 1957 to the Valley Tonga people of Zambia and to the upheaval in their lives posed by the building of the Kariba Dam. At the time when I was invited to make a human biological study of these people, Desmond was the curator, and his wife Betty, the assistant curator, of the Rhodes-Livingstone Museum in Livingstone. This sunny and pleasing town was close to the spectacular Victoria Falls on the left (north) bank of the Zambezi River in Zambia (then still called Northern Rhodesia).

Clark had been appointed to the curatorship of the Museum when he was a stripling of 21. In the course of his university education, he had fallen under the spell of that great Cambridge teacher, Miles Burkitt (see Chapter 11). Those were the days when an academic's reputation did not rest upon a Ph.D. degree and Burkitt went through his remarkable career without any doctorate. Despite this, he contributed more dedicated archaeologists to remote corners of the world, and especially Africa, than did anyone else I can recall. Desmond Clark was one of his intellectual progeny. It is fascinating to recollect that Burkitt's classical text, *S. Africa's Past in Stone and Paint* (1928), held sway for 30 years until it was superseded by Clark's book on *The Prehistory of Southern Africa* in 1959.

The Second World War interrupted Clark's 23 years in charge of the Rhodes-Livingstone Museum. He joined up and saw service in Ethiopia, Madagascar and the Somalilands from 1941–46. Even during those years he kept his eyes peeled on the soil of Africa and its prehistoric treasures. His work in the Somalilands and Ethiopia at this time laid the foundations for one of his early books, *The Prehistoric Cultures of the Horn of Africa* (1954).

In 1961 Clark was recruited by Sherry Washburn to the University of California, Berkeley, where he remained for 40 years. Now less isolated than in the depths of tropical Africa, his message went far and wide from

Berkeley. All over the world there were scholars and students who owed a debt to Clark. They had cause to be grateful to him for help, encouragement, advice, example, hospitality, and his characteristically unselfish attention to their needs. Clark's touch was evident from Korea to Zambia, Morocco to Madhya Pradesh, Malawi to Niger, the Awash to the Zambezi, Berkeley to Bryn Mawr, and Stanford to Sudan.

The world was his playground and of no one may it more truthfully be said that his life and his manner fulfilled the status of a world scholar. He illustrated that science was above nationality, partisanship and insularity. International savants such as Desmond Clark are citizens of no country, place or region: he was verily a citizen of the world.

Like my late father and myself, Clark was a devotee of Gilbert and Sullivan. Knowing this, when I was invited to deliver a toast to Clark at an international conference in his honour to celebrate his 70th birthday, I could not resist delivering half of my libation in verse based on *Pirates of Penzance*, *Patience*, *Iolanthe*, *Mikado* and *Gondoliers*. Nobody joined in the choruses more lustily than Desmond, and I have had to exert Herculean self-control to resist quoting some of these stanzas here!

The Annual Winter Schools that Clark introduced in the Rhodes-Livingstone Museum when there was as yet no university in the territory, spread the message among professionals and amateurs. I recall these with nostalgia, as they brought me from Johannesburg in the 1950s on many happy and convivial visits as one of Clark's team of Winter School lecturers. I usually remained for a week and delivered a course of four or five lectures on physical anthropology and what we had not yet started to call palaeoanthropology; essentially, on the people behind the tools. Desmond and Betty were my hosts during these stays.

These Winter Schools marked the start of our 50-year friendship that endured until Desmond's death at 85 years of age in 2002. That is a long time for a friendship to last – unalloyed by a single cross word, a single heated difference of opinion – in the world of the twentieth and twenty-first centuries when friendly consensus and agreeing to differ seem to have given way to edgy confrontation.

The groves of academe have, sadly, not been exempt from the decline of human relations. Conscious of this, I rejoice at the worldwide camaraderie of scholars and I exult in the links of loyal comradeship between Desmond and Betty Clark and myself.

IF OUR FRIENDSHIP had started with the Museum's Winter Schools, it was cemented when I was invited in 1957–58 to participate in the Gwembe Valley Survey organised by Desmond Clark and H.A. Fosbrooke, the director of the Rhodes-Livingstone Foundation in Lusaka.

The construction of the Kariba Dam across the Zambezi River in the 1950s was an inter-territorial project. Zambia lay on the left or northern bank and Zimbabwe on the right or southern bank. At that time the three territories of Zambia, Malawi and Zimbabwe were united in the Federation of Central Africa, a geopolitical consideration that made it easier for the planning of Kariba Dam. The planners realised that a vast area of the Gwembe Valley in Zambia's Southern Province would be inundated by the rising waters of the dam. The consequence was that ecological, geological and archaeological features of the valley would be buried beneath about 100 metres of water.

Even more cogently, the village homes of tens of thousands of Tonga people would be flooded and covered up. It was estimated that there were some 34 000 people on the Zambian side and just over 20 000 on the Zimbabwean side. Resettlement of the people to the higher-lying slopes of the deep Gwembe Valley was planned by the two territorial governments, but the people took some persuading. All their lives they had lived on the banks of the Zambezi and they knew the river's moods and phases. It was inconceivable to them that the river could be changed as the government officials were telling them. On the Zambian side the authorities relied largely on persuasion and voluntarism, although there were some pockets of strong resistance. On the Zimbabwean side it seems that a greater coercive element was applied.

Faced with this drowning of an immense valley and all it contained, a multidisciplinary research survey of the Gwembe Valley was conceived by Clark and Fosbrooke. A team of international scholars was recruited to study different aspects of the area and its inhabitants that would fall under the future water level.

Elizabeth Colson and Thayer Scudder undertook the social anthropological and ecological programmes; Barry Reynolds studied the material culture of the living peoples; Desmond Clark, helped at the time by Brian

Fagan and Ray Inskeep, carried out the archaeological survey. Others who participated from time to time were John Blacking on ethnomusicology and Clayton Holliday.

I was invited to undertake a human biological study of samples of the Valley Tonga population, as well as of Plateau Tonga as an outgroup for comparison. My assistants were several Medical B.Sc. and Honours graduates of Wits whom I had trained in anthropometric and other human biological techniques. On the 1957 expedition they included Lawrence Geffen and Martin Bobrow, and on the 1958 expedition Richard C.J. van Hoogstraten, Adrian Roy Percival Martin and Dawood Mangera. The presence of Dawood made this effectively the first 'multiracial' expedition from Wits. Elizabeth, the daughter of Desmond and Betty Clark, dubbed us 'Tobias and his angels' and Betty regularly enquired after my angels for years afterwards.

Our multidisciplinary undertakings had the amusing consequence that at times as many as half a dozen scholars were at work in the valley in 1957–58. I remember one occasion when no fewer than four vehicles intersected paths in the midst of the Gwembe Valley. A mini-symposium ensued (and you might remember that the literal meaning of the word symposium is 'drinking-party'). The disciplines represented were social anthropology, ecology, archaeology and human biology.

A special feature of the Zambian resettlement plan was that each village was moved intact to its new location above the future waterline and sometimes even along the same tributaries. My concept from the beginning was to do a before-and-after study rather than a once-off survey. I proposed to follow up the same sets of villagers some time after they had been resettled. The point of this was that conditions in the Gwembe Valley before flooding were appalling: under-nutrition was rife, malaria and bilharzia (schistosomiasis) were present everywhere, endemic goitre was frequently encountered in certain areas, sleeping sickness (trypanosomiasis) was occasional and the cattle version of sleeping-sickness, nagana, had wiped out most cattle in parts of the valley. Rampaging hippopotami were a regular threat to the crops of those Tonga living on the bank of the Zambezi. The loud cries of the people and their beating on metal drums and lids, often as we were settling into our sleeping bags and adjusting our mosquito nets, was a signal that hippos were abroad and making a nuisance of themselves.

With all these environmental threats, it is no wonder that the physical state of the Valley Tonga before resettlement was well below par. It was the governments' plan to stock the new 'Lake Kariba' with edible species of fish and to persuade the people to incorporate fish into the food culture of the resettled villagers. If this worked, it was surmised, the state of nutrition of the people would be substantially improved by the ingestion of abundant fish protein. For this reason it was my long-range plan to return to the Gwembe after a generation or two, to see whether physical improvement was detectable by our human biological techniques. So the 'before' study was made in 1957–58. Although I published numerous articles on the Tonga adult data between 1966 and 1985, various factors delayed the 'after' study, until an American student-collaborator, Rhonda M. Gillett, embarked on follow-up studies in the 1990s. Her work, partly jointly with me and partly on her own, is meticulously giving effect to precisely the sort of follow-up research I had envisaged back in the 1950s. The lapse of time between the 'before' and 'after' studies was nearly 40 years. This has been long enough for any nutritional improvement to have become evident. It is long enough also for secular trends to have declared their presence (see Chapter 24), and this factor may complexify Rhonda's and my analyses.

The analysis of my 1957–58 data by Rhonda Gillett started with the children of the Valley Tonga and with her data for children in the Plateau town of Choma. The results were published by us in 2002. They suggested, as we had observed clinically, that children in the Gwembe Valley at the time of resettlement (1957–58) were nutritionally compromised. High percentages of them suffered from chronic protein energy malnutrition. Some children were underweight; others were stunted in stature; whilst many were both stunted and underweight. Dr Gillett, who is at the University of Arizona in Tucson, is now comparing our 1957–58 measurements with those she obtained in the 1990s to test our prediction that there might be an improvement in nutritional status, either from the incorporation of fish protein into the diet or for other reasons.

OUR FIELDWORK WAS carried out in the mild winter months of July and August in 1957 and 1958, while other members of the team gathered genetical data in the early 1960s. We studied some 400 adult males and over 600 children. Our subjects were drawn from Valley Tonga who were to be resettled. The data for these were to be compared with data for two control groups. The first group comprised Valley Tonga living at that time just beyond the future waterline and close to the planned resettlement areas; whilst the second control group consisted of Plateau Tonga. Close on 200 observations were made on each adult subject. These embraced physical anthropological, genetical, nutritional, medical and somatotypic (physique) features. Among the special features of the survey, we took fingerprint records, observed morphology and anatomical variants, and the incidence and patterns of colour-blindness. For most of the 200 characteristics, no data had been previously collected in tropical southern Africa. We visited a number of schools and were able to study more than 600 children. On each child we measured height and weight, recorded the stage of emergence of the teeth, and made genetic observations.

From an analysis of all of the data, I hoped that we would be able to draw conclusions about the physical make-up, the physique and nutritional status, and the genetical composition of the Tonga people in the area studied, especially those who were to be affected by the resettlement.

We worked from three or four base camps that were set up from west (nearer the Victoria Falls) to east (nearer the dam wall). Each day, we packed up our equipment, water and *padkos*, and set out for one or two villages. Once the elders had given their permission we spent the entire day working on and interacting with the villagers, adult males and especially the children. At the end of each day, we packed up all our goods and climbed aboard our five-ton Bedford truck for the journey to our home base. Time and again, as we were about to leave, scores of the little ones came running alongside and begged for a ride. I could not resist their entreaties: twenty or thirty would climb aboard and start singing. These were magical twilight moments and we took them for a short distance. For their safety and our own concern not to get lost, for the roads were nothing more than tracks through thick bush, we thanked them for their songs and asked them to climb down. But thanking them for the songs was equivalent to an (unspoken) request for another song! I have always loved choral music and especially the abandon of African choirs. Those evening rituals were among the happiest times on the hot, dry, dusty days filled with work.

One of our camps was on the shore of the Zambezi River. We had seen hippos in the wide river and on two nights they came right through our camp, leaving their pug-marks perilously close to our tents and fireplace. Crocodiles were also a deterrent to our cooling off and having our nightly wash in the river. We had no firearms and one of us would stand guard with a lamp to shoo away any hungry or venturesome reptiles – a thought that added a tingling zest to our ablutions.

Opposite our camp was an island with a winter pumpkin crop. As our supplies were running low, we procured a pumpkin or two. For days, our cook who had come with us from Livingstone, dished up roast pumpkin, fried pumpkin, pumpkin fritters, boiled pumpkin, cold pumpkin, and mashed pumpkin with local field relishes. It had recently been demonstrated that pumpkin contains goitrogenic (goitre-forming) substances, and we were in an area where goitre was endemic – we had seen a number of subjects with enlarged thyroids. Consequently, our excessive ingestion of pumpkin gave me a little anxiety. I took to measuring the circumference of the neck of my two 'angels' every morning at the breakfast table, to see if I could detect any signs of enlargement of the thyroid gland. Happily none of us showed thyroid enlargement, but we abandoned pumpkin as soon as we were able to get some other foodstuffs.

I realised only 34 years later how much the students' participation in the Gwembe Valley expedition meant to them when I read this passage by Laurie Geffen in 1991:

> I count those several weeks spent in Tobias's company, engaged side by side with him in the field work he loves so much, as amongst my most precious memories. The lessons I learnt from this great man about the discipline and the pleasures of scientific discovery left indelible impressions.
>
> I still remember, to my shame, sulking in my tent because he would not allow me to keep as a memento even a single stone arrow head I had picked up on an ancient river bank, despite the fact that the gravel bed was littered with thousands of (to me) identical artefacts. Tobias insisted that it be properly labelled and put in a museum collection; the object did not matter but the principle did!

> I remember also the crowds of pot-bellied Tongan children who followed him and fell about with hilarity when he peered into their mouths and guessed their ages from their erupting dentition. He was later to amaze my own children with this party trick on his visit to Adelaide in 1979 and almost launched them on careers in dentistry![10]

By the time I assumed the chair of anatomy in succession to Raymond Dart in 1959, I had become well acquainted with two big branches of living southern African humanity. My Kalahari adventures had introduced me to the Khoe-San people, while the Gwembe-Kariba venture had opened my eyes to a major branch of the Bantu-speaking Africans, namely the Tonga. One was a population that had been steeped in the hunter-gatherer tradition, although its dependence on this mode of subsistence was diminishing; the other branch of African humanity lived by agriculture and, where their cattle had not been wiped out by the ravages of nineteenth-century raiders and of trypanosome-infested tsetse flies, the pastoral mode of life was added.

When I looked at the short, light-skinned and relatively hairless, small-boned and delicately built San and then at the somewhat bigger and definitely darker-skinned, rather more robustly constructed Tonga, I could not help wondering how much of these distinctions were to be laid at the door of the environmental differences, the varied ecologies in which they had been reared and to which they had become adapted, and how much was owing to genetic variations.

This was to be an abiding question during all of my ensuing years. I was living in a country and a politically dominant society, that of apartheid, in which virtually all differences among human beings were ascribed to race. This obsessional attitude applied not only to physical features and genetic traits. It was also wantonly invoked by cabinet ministers and ordinary people to 'explain' psychological, behavioural, intellectual, educational, ethical and attitudinal characteristics. Variations in these, however poorly proven, were blamed on race. The San and the Tonga provided striking examples of the opposite standpoint: no one could escape the overwhelming impact that ecology – environment in the broadest sense – played in their respective societies.

24 | Growing up or growing down?

Of the many features of the San that set me pondering, their short stature was most noteworthy. Had they always been short? Was it in their genes that their shortness was founded? Was there anything in the claims of earlier investigators that the shortest San were the 'purest'? I knew from my reading of earlier reports that some of the nineteenth-century scholars had accepted this idea. For them, taller San implied racial admixture with other, taller peoples. So some of the early anthropologists selected only the shortest Bushmen for study, discarding the rest as 'impure'. If the San had had a better diet and water supply, would they on average be taller? Why are they of shorter average height than their black African neighbours?

To seek answers to these perplexing queries, I collected data for as many other African population groups as I could, either by direct personal measurement or from published accounts. In this quest I was aided by my team from the Anatomy Department, including Trefor Jenkins after he joined me in 1963. We collected data on adult males from many southern African populations and chiefdoms. These data were to fill a big gap in our knowledge of stature and other features of African peoples from the southern sub-tropical and tropical populations. It struck me forcibly, once more, how much taller most of these people were than the San Bushmen.

A TANTALISING FEATURE had been revealed in human biological studies in Europe and North America shortly before, and especially after, the Second World War. This was a tendency for children to mature more rapidly and earlier. The trend showed itself essentially in two ways. In height and weight, children had become appreciably larger at each age than children of similar ages had formerly been. Secondly, the changes at puberty, especially the growth spurt and menarche (the onset of menstruation), were occurring at progressively younger ages. Hence most studies on the secular trend in humans were based on statistical data on the age at menarche and on growth in stature and mass.

The term 'secular' used in this biological sense has nothing to do with 'secular' as used in theological discourse, and means lasting for an age, or for a hundred years, or for a very long period of time, as distinct from short-term or periodical changes. It is secular change as distinct from evolutionary or genetically determined change.

'Secular change' gradually acquired a wider use when it was applied to adult features, especially to average height. When earlier and later measurements on adults of the same population were compared, it was found that people in many parts were growing taller. The young gentlemen of Harvard University provided an especially glaring example. Their statures had been recorded since the nineteenth century and in many cases members of the same family were found to have increased by about half an inch (or 13 millimetres) per generation. This had gone on for successive generations, so that the mean height of Harvard students could be shown to have increased progressively. I remember discussing this with Al Damon at Harvard in 1956. Surely, we were saying, this trend cannot go on indefinitely? There must be an upper limit set by the genes. We haven't found a limit, said Al. About a year later, I wrote to him and asked what the latest set of analyses had shown: were the statures still 'going up'? His reply was historic. The latest data showed that in those families that had been affluent and presumably well nourished for the longest time, the new generation of Harvard students showed no increase over the last. 'Three cheers!' wrote Damon, 'The genes do have an effect!'

What seemed to be happening was that the genes set a limit for adult stature, but it is not the inheritance of a specific height, but of a range of possible heights. Then for each individual the diet and perhaps other features of the environment select the stature within this gene-determined range. By way of example, if you were born into a family in which tallness runs, your genes would not dictate that you would be, for instance, 2.1 metres tall, but rather they would stipulate a range of tallness, such as about 1.9 to 2.2 metres. Diet, exercise, rest and recreation would be expected to make the fine adjustment of your actual adult stature within this range.

Up to about 1960, the secular tendency towards growing up had been positively identified for a number of European, North American and Hong Kong populations. It was assumed that this was a general trend for all modern humans, but I found that no single published study of any African peoples had revealed that secular changes had occurred in this continent. (There was one exception to this that became known to me only later, when I obtained a copy of the unpublished M.D. thesis of Dr Sidney L. Kark, formerly of the Wits Medical School, later of the University of Natal and then of the Hebrew University, Jerusalem. He and his wife Emily were pioneers of community health in South Africa and Israel. From his data on Zulu people, Kark found indirect evidence that they had shown secular changes, although he did not relate his results to either the concept or the term, secular trend.)

Armed with my San data from the Kalahari, I sought reliable reports on San statures from earlier publications. For the San groups I had studied after the Second World War, I found published results dating from the period between the world wars, and from the period before the First World War.

It was astonishing to find that the San I had measured in the 1950s were on average taller than members of the same groups had previously been. My article of 1962 was the first to publish evidence that any African peoples had shown secular changes. What was so surprising was that, not only was this the first African population to show a secular trend, it was the first time that any population of hunters and foragers (sometimes called a Fourth World community) had shown this trend. Previously, secular changes had been observed in populations in industrialised countries (commonly spoken of as First World and also Second World communities).

James ('Jim') Tanner of the Institute of Child Health at the University of London was one of the world's leading authorities on human growth and development. His book, *Growth at Adolescence* (1955), was a source of inspiration to growth scholars, including myself, in many parts of the world. He encouraged me greatly at a time when I was the only person in Africa working on secular trend problems. In 1962 Jim Tanner reviewed the various causes that had been proposed to explain the secular trend towards faster growth and bigger adult bodily size. The four principal hypotheses were that it was due to nutritional improvement, other environmental factors aside from diet, a rise in world temperature, and a possible genetic cause.

With these competing hypotheses, Tanner was forced to conclude that 'Nobody knows for certain why the secular trend has occurred'. I added a different nuance when I stated that there might not be a single cause or causes operating in all populations; different causes might have applied to varied populations living under diverse circumstances.

Of the various factors that were proposed as causing secular changes, it seemed to me that in the San the most likely cause of their secular trend was the environment, for there was evidence of slight environmental amelioration. The signs included the laying down of boreholes by government officers, and the distribution of famine relief to the San by the government of what was then Bechuanaland.

If the most affluent and presumably well-nourished people and the most disadvantaged populations on earth both showed secular changes towards greater adult height, it would seem to follow that the great majority of the world's people falling between these extremes (the so-called Third World populations) would also show similar secular changes. In other words, my demonstration of the secular trend in the San appeared for a time to strengthen the case for all of living humanity to be showing the secular trend towards taller stature – at least until I subsequently overturned this notion.

UP TO THE 1960s almost all scholars working in this field spoke and wrote about *the* secular trend and it was commonly thought of as one-way traffic. If a population showed secular changes, they were towards faster and earlier growth, and towards taller adult stature.

I was not satisfied with this assumption. Here in Africa were hundreds of millions of people living agrarian or pastoral lives in societies that were non-industrial or pre-industrial. I was determined to find out whether samples of these peoples were indeed growing taller.

As soon as I could gather together earlier and later measurements on some of these populations, I found a striking contrast with my results on the San. There was almost no sign that African people were growing taller. Instead, I was amazed to find glimmerings of a secular trend but in a reverse direction. My first results were on South African Bantu-speaking people who appeared to be growing shorter! Could this 'growing down' effect, I wondered, be an insidious consequence of the apartheid policy that prevailed and was indeed intensifying at the time? However, I found samples from other parts of Africa, and they also showed trends opposite to the usual secular trend. A few results I collected from Brazil and northern India similarly showed a reversal of the usual secular trend. Wherever Third World populations were examined, they revealed either an absence of secular changes, or reversed secular changes.

I listed 22 features of newborn babies, immature individuals from the neonatal to the young adult stages, and adults. It emerged that until that time, the term secular trend was applied to changes in some or all of these 22 traits *in one direction only*. That is, the term was considered to cover an *acceleration* of the growth process and the attainment of *larger* bodily dimensions, whether at birth, at any stage in the growth process, or in adulthood. Now, with the new data I had found in non-industrialised populations, to which were soon added results by others, it was no longer acceptable to speak of '*the* secular trend'. I proposed that human biologists should use the terms 'positive secular trend', 'absent secular trend' and 'negative secular trend'. In particular, while my publications of the 1960s and 70s clearly showed evidence for reversed and absent secular trends, the concept of the *negative secular trend* was firmly established in my article published in the *Journal of Human Evolution* in 1985.

After I had published a number of papers on what I at first called a reversed secular trend, I received a letter from Dr Edeltraud Ludwig of Saarbrücken University, Germany. She told me about a Doctor of Philosophy thesis by Georg Kenntner of the University of Saarland, Karlsruhe. Kenntner used earlier and later data for populations in some 38 countries or areas. For 24 countries or areas there had been an increase in average adult stature in the previous 50 to 100 years; for 12 stature had remained unchanged over that period; and for 2 it had declined. The 12 territories in which mean adult stature remained unchanged included Côte d'Ivoire (Ivory Coast), Upper Volta, Senegal and Sudan. The 24 countries for which an increase was recorded did not include a single African territory. Although Kenntner did not relate his interesting results to the secular phenomena, his work confirmed that absent or negative secular trends were widespread in Africa.

When I look back on those years when I was busily measuring people from all over southern Africa, it strikes me as remarkable that such an extensive output of facts and concepts can flow from so small an input of technical expertise – the simple measure of bodily height.

FEMALES ARE, AS everyone knows, the stronger sex. My cousins, Olga, Joy and June, fully agree. To help males get used to this idea, my old and very literate friend, Ashley Montagu of New Jersey, in 1953, wrote a book called *The Natural Superiority of Women* – but you should really see the new expanded edition of 1968! If only Ashley had asked me, I could have given him some new ammunition. For the moment I started analysing my data for female San, as well as those of the males, the greater resistance and resilience of the ladies to adverse conditions became immediately apparent.

What I have said so far has been based on measurements of adult males. When I enriched my database by including the heights of females, more unexpected and exciting results flowed forth.

Keeping to the adults, I first examined the differences between the average statures of adult males and females. The technical term for this is sexual dimorphism. Again I was to stand agape at my results from two such simple measures as the average stature of adult males and females.

In 1962, I demonstrated that the positive secular trend among the San had affected the two sexes but to different degrees. For each group of Bushmen, the increase in average stature was roughly twice as great in males as in females. As a result I was able to claim in 1962 that with a positive secular trend of adult stature, 'there is an increase in sexual dimorphism in stature'. This was true of the absolute differences between the average heights of males and females. I found it was true also when I expressed the absolute height differences as a percentage of the average adult heights, a ratio we call relative sexual dimorphism

The San, it should be remembered, were the only African population to show a positive secular trend. So we can say that as the heights of the San had grown bigger, males showed a larger increase than females. Since in any human population, males are on average taller than females, this difference for recent San was greater than it had been between the world wars and before the First World War. If it was as I believe good conditions that led to the positive secular trend, we may say that under such conditions males soared upwards more than females did. This was true for all three San groups that I studied in the 1950s, namely, Northern, Central and Southern Bushmen. Hence as San heights had increased, so had San sexual dimorphism increased.

When, however, a large series of non-San, sub-Saharan Africans was studied something different came to light. Jean Hiernaux in 1968 reviewed the available data on sexual dimorphism of stature in European and sub-Saharan African populations. However, he included only two southern African Bantu-speaking populations, the Nyungwe of Mozambique and the Venda of the northern Transvaal (now Limpopo Province). I re-examined all of Hiernaux's data, supplementing them with additional data I had assembled. The resulting values for 49 sub-Saharan African populations were compared with those Hiernaux had put together for 41 European populations. My analysis of the larger African samples and the same 41 European samples revealed that sexual dimorphism in sub-Saharan Africa was smaller than it is in Europe. In this, both Hiernaux and I agreed.

MERE FACTS AND their statistical analysis are only the beginning of scientific research. Science is different from mere stamp-collecting (and I don't mean to knock philately – far be it from me as a collector for the past 70 years). Once the facts have been obtained, you try to see what they mean and to develop an hypothesis to explain them.

Here we have data for 90 modern human populations. Those of Europeans have higher, and those of Africans lower, sexual dimorphism. It would be facile to dismiss these differences as the effect of genetic differences, and we have no evidence that differing packets of genes are responsible for the amount of sexual dimorphism. There is certainly a well-attested environmental factor that would seem to influence the degree of sexual dimorphism. African populations in general live in less favourable environmental conditions than European populations although there are exceptions. Herein may lie the main cause for the lower sexual dimorphism of stature in sub-Saharan African peoples.

What is the usefulness of this sort of study, you may be wondering? It is an interesting corollary that this fact may give us an index of environmental, including nutritional, amelioration. There are today deplorable levels of under-nutrition, malnutrition, marasmus and kwashiorkor in Africa. As living conditions in Africa improve, and improve they must, we may expect the populations to show an increase not only in average stature, but also an increase in the degree of sexual dimorphism. This is a deduction. So far it is supported only by the San data, which reveal the positive secular trend and with it an increase in sexual dimorphism.

Perhaps the San may prove to have pointed the way to a concept of some heuristic value in further studies on human populations and their adaptability.

I'd like to add one last thought on sex differences of the growth rate curves of boys and girls. It is well known that the comparison of growth rates among juveniles provides a useful yardstick of the nutritional status of a community. In 1970 I showed that some reliance may be placed upon sexual dimorphism of growth curves. We know that girls are ahead of boys from puberty onwards and they start their pubertal growth spurt earlier than boys do. When this happens girls become taller than boys of the same age, always a trifle embarrassing in co-educational schools and in families with boy-girl twins! We call this feature of the growth curves crossing over. Then, after several years, the boys belatedly start their growth spurt and their heights overpass those of girls of the same age. The female growth curve is said to cross back. I gave the name 'the period of female ascendancy' to the age period between crossing over and crossing back. My data have shown that the period of female ascendancy starts later and ends later in poorly nourished populations. Moreover, the period of female ascendancy is of longer duration under poor conditions; that is, the difference between the ages of crossing over and crossing back is greater. Conversely, I pointed out, environmental betterment may be accompanied by a decrease in the duration of the phase of female ascendancy in juveniles.

In closing this chapter on secular trends, let me share with you a question that has been tickling my mental faculty for some time: are secular trends of any evolutionary significance? I have carefully distinguished between secular change and evolutionary change, and they are diametrically different concepts. The thought has often crossed my mind that populations may vary in their ability to make secular adjustments. We have no evidence for this, to the best of my knowledge, so it is speculative. But if one population did possess a stronger competence for secular adjustment, would it not, in the face of deteriorating environmental conditions, be at an advantage, as compared with a population whose development was more rigidly canalised, less flexible and less liable to secular adjustments? I don't know the answer but I sometimes let my mind wander over such matters . . . Then I answer my own question: yes, it must surely be so! It has been said that 'Nothing in biology makes sense except in the light of evolution.'[11] Does this not apply also to secular tendencies? Now it remains for me to find a Ph.D. student or post-doctoral fellow to explore this revolutionary notion and disprove it . . . or otherwise.

25 A reluctant professor

I did not want to become a professor or a head of department. This issue first reared its head in 1955 when I was persuaded by Dart to apply against my better judgement for the chair of anatomy at Cape Town, which was due to fall vacant at the end of 1955.

On 28 October 1955, I recorded in my journal: 'I don't really want a chair, but at the same time, feel that things are moving as if by predetermination in that direction. I feel a little at the mercy of inexorability or inevitability ... Further, I am ... rather acutely conscious, of a sense of inadequacy before this inevitability.'

Then I dismissed the matter from my mind because of sheer busyness. To my great relief, Lawrence Wells, a respected and loved former teacher of mine who was currently a reader in the Anatomy Department at the University of Edinburgh, was offered the chair. He became professor of anatomy at the University of Cape Town in 1956.

The sentiments I inscribed in my journal in October 1955 were still with me three years later. I loved my students and delighted in my role as a teacher. From the beginning I had found that I had a penchant and a natural bent for teaching. I knew that if I were to become a professor and head of department, I should have far less time to spend with the students and less by far to engage in instructing them. I am grateful to my friend, the inspired artist, Nils Burwitz, for introducing me to George Steiner's book, *Errata: An Examined Life*. There was a passage that reminded Nils of me and my passion for teaching: '... I knew now that I could invite others into meaning. It was a fateful discovery. From that night on, the Sirens of teaching and interpretation sang for me.'[12]

How much of that passion would I lose if I became a professor and a head of department?

My strong enthusiasm for the spirit of inquiry also engulfed me. How much of this love for my research would the appointment permit me to indulge? The point was vividly illustrated to me by Theodore ('Teddy') Gillman (Joe Gillman's younger brother). When he heard that I had been appointed to Dart's chair, he diluted his felicitation by mordantly saying, 'You, with your dozen publications a year – you must not expect to publish anything during your first five years in the chair.' I was shocked and stunned at this gloomy prospect. I had feared that there would be a decline in my research output, but not on such a scale as Teddy was now prognosticating. Stung by his cynical 'good wishes', I overcompensated in my first five years in the chair and published more works per annum than before! It may have cost me sleep and social life and recreational activities, but I disproved Teddy's prediction and somewhat restored my peace of mind.

A third factor that distinguishes between a head of department and other staff members is the much bigger load of administrative duties with which the chair is saddled. Most of these duties are in the running of one's own department; others relate to the Faculty Boards – for at that time the Anatomy Department belonged to three faculties: Medicine, Dentistry and Science. In the event, when I became a Senate representative on the University Council, and of its Executive, Staffing and Promotions Committee and sundry others, my secretary tallied up that I had to attend about 65 meetings a month! Although I seemed to have a flair for committee meetings, the inordinate number of hours that I was obliged to devote to these, robbed me of much of my available time for students and research.

With this three-pronged potential assault on my two main loves as a citizen of academia, and insufficient ambition to outweigh it, it is small wonder that I made no attempt to apply for Dart's job at Wits. He was away on long leave and I was acting head of department in the last months of 1958. Dart arrived back a day or two before the closing date for applications for his chair. When he found out that I had not applied, he called me to his office and scolded me severely, with a rich interlarding of expressive Australian adjectives and sundry other expletives. 'What the ... do you think I have been grooming you for all these years?' he ejaculated. Then he made me sit down and write out my application without delay.

It was just over a month before my 33rd birthday, when the Selection Committee called me for an interview. I had a luxuriant red beard – which had grown during my latest excursion, to study the Tonga people in the Southern Province of Zambia (see Chapter 23). The chairperson of the Selection Committee, Professor I.D. MacCrone, was surprised to see a bearded apparition enter the room. He teased me wryly: 'We know why you're sporting a beard – to convey to the Selection Committee the impression that you are older and more mature than you look without one!' It was a searching but not unpleasant interview.

As I walked down the passage afterwards, I saw a Cape Town friend walking in the opposite direction. Naïvely I asked, 'What are you doing here?' He replied, 'The same as what you are doing!' Only then did I realise that he, too, was a candidate for the Wits chair.

ONE HUNDRED AND eleven days later, I found myself in the professorial office, called upon to lead a department that was depleted of staff, with almost no research funds, no departmental library, no journals club, and a storage museum but no teaching museum. In honour of my being appointed as the youngest professor at Wits, my cousins Ben and Zema Mordecai (Ben was the twin son of my father's sister and hence my first cousin) arranged a memorable party at their small suburban home in Johannesburg. Two or three hundred people were present. My father and stepfather were sitting chatting on a sofa, when some of Dad's Johannesburg friends came up to him. He had been celebrating rather freely and on such occasions enjoyed his Scotch whisky. A little unsteadily he rose to his feet and gesturing towards Bert Norden announced, 'Allow me to introduce my wife's husband.' I still enjoy a chuckle at the memory of his words and the outburst of hilarity that greeted them.

The prospect of following in the footsteps of so gigantic a personality as Raymond Dart was daunting. I remembered the novel, historical-in-prospect, that Arthur Keppel-Jones published when Jan Smuts's time was getting near: it was called *When Smuts Goes* (1947). There, too, a successor was called on to walk in the footsteps of another giant of a man. He could not do it on his own: teamwork and consultation were of the essence.

So minded, I called a meeting of all my staff. I freely acknowledged my juniority of age and that some of those in the room were older than I was. I declared that I proposed to hold regular staff meetings (something new to the department), whereby I could consult those older and wiser than me, as well as the young and innocent, out of whose mouths now and again wisdom might come forth. They took the point and received my ideas graciously and even with a burst of enthusiasm.

I stated also that I proposed to appoint as assistant lecturers several surgical aspirants who wished to learn anatomy by teaching it, in order to equip themselves for the Primary Examinations leading to the Fellowship of the College of Surgeons of South Africa (the postgraduate qualification of surgeon). Dart had appointed a few people in this category in the last years of his reign. I set about increasing the numbers and giving them a bigger role in teaching at the cadaver side in the dissection halls, as well as in other functions. Two teachers in the department offered special courses of clinical anatomy and applied embryology and teratology (the study of congenital malformations) to these surgical aspirants.

One of my greatest joys was the course I developed in embryology and teratology over the years. This was open to staff members of all manner of backgrounds, although intended primarily for those younger doctors who were bent on surgery. The head of the department of surgery, Professor Daniel J. ('Sonny') du Plessis, regularly advised those who consulted him about careers in surgery that the best approach was to spend a year with me in the Anatomy Department. The numbers applying for these jobs each year increased far beyond the five, six or seven vacancies we had for such posts. Over the next decades, some 150 surgeons received their anatomical grounding in the Anatomy Department. These numbers went well beyond what I had envisaged at that first staff meeting.

Having seen the admirable self-teach facilities in Solly Zuckerman's Anatomy Department at Birmingham, England, I told the staff that I envisaged converting our antiquated storage 'museum' into a proper self-teach museum on similar lines. Eventually, I convinced the university and in 1968 a superb new two-tier Hunterian Museum was erected over the lower dissection hall, once its foundations had been strengthened to support a two-storey structure. The study museum was more than just a collection of exhibits: it was arranged to be a facility for students and others to learn. The senior technologist, Roland Klomfass, prepared some extraordinarily fine self-explanatory mobile specimens.

I also told the staff I wanted to start a departmental library – we had none before – and this took off almost immediately.

Thus, by a series of developments of this nature, the staff and I gradually built up a worthy department.

An unexpected by-product of the assistant lecturer scheme was that several of the 'table doctors' undertook research projects. A few even registered for Ph.D. degrees with me (such as Errol Levine, Brian Wolfowitz and David Ricklan) and successfully completed them before returning to their medical specialities. These people formed part of one cohort of my research students. There were, of course, many others who came to do research in physical anthropology, mostly palaeo-anthropology, all working primarily on fossil bones and teeth. Yet others were drawn to doctoral studies on the living.

So a part of the building up of the Wits Anatomy Department lay in the vigorous development of research schools, including both Master's and doctoral students and post-doctoral fellows.

From the beginning I did not wish to craft a monolithic department in which a single major research thrust would predominate. I knew that very little research was being carried out in the few other anatomy departments in southern Africa at that stage. I felt a strong duty to South African academia as a whole, therefore, to keep a number of areas alive and to encourage their development. The wide range of topics reflected in the publications, dissertations and theses in the Wits Anatomy Department mirrors the success of this objective over the next 30 or 40 years.

One major focus was that of embryology, especially experimental embryology, conceptually a most advanced branch of morphology. Another was the study of the nervous system; its structure and comparative morphology had long been an active branch of the department. Genetics, particularly cytogenetics (the study of chromosomes), was a further active focus of our work, and already two higher degrees had been awarded in chromosome researches, Sydney Brenner's M.Sc. and my Ph.D. Some active campaigning on the part of myself and the head of the Department of Pathology, Professor B.J.P. ('Bunny') Becker, saw the installation of the Wits Medical School's first electron microscope in 1966. Undoubtedly the most vigorous and productive of our research entities and the first to become a full-fledged research unit, was that of physical anthropology, including palaeo-anthropology (the study of hominid fossils and human evolution).

These five blocks of emphasis were not the only ones that, at one time or another, kept me and some members of my staff busy. The multiplicity of the Anatomy Department's interests was formally recognised by Wits when, in 1980/1982, it was awarded a Certificate for Outstanding Achievement in Research and Scholarship of Excellence.

A GROWING AWARENESS of the research fields in anatomy helped to generate the Anatomical Society of Southern Africa in May 1969. At this Society's annual meetings, for a number of years half or more of all the papers presented emanated from the Wits Anatomy Department alone. This was all very well as a measure of the success of the Wits department, but from the time when I first set up the Society, it had been my hope and ambition that it would catalyse researches at our sister departments at other universities. Fortunately, that is what happened. Year by year, I saw with pleasure that more and more research was being reported from other departments. Inevitably, the establishment of the Society provided a ferment for all of the departments of anatomy in southern Africa. From the beginning, we used the term 'Southern Africa' – because medical schools were springing up in Zimbabwe, Zambia, Angola and Mozambique. It was our fervent hope that our anatomy colleagues in this broader geographical region would become part of the Society. This was plausible and tolerably successful, at least until the rise of the anti-apartheid academic boycott chilled these relationships.

The Society still exists and flourishes. Under the new democratic South Africa, international linkages have become easier and moves are afoot to forge closer links with schools further afield – as in East, Central, West and North Africa. Although my own pressures have led me reluctantly to play a less active part in the Society in later years, my interest in, and devotion to, the Society's ideals have remained as strong as ever. My hope that the Society would be able to start its own periodical have not yet borne fruit, but at least a newsletter has come into being. Under the active editorship of Beverly Kramer, the Anatomical Society newsletter developed a sparkle. She had been a professor of anatomy in the Wits Dental School (which for several decades had its own basic medical science departments) and recently in the reunited School of Anatomical Sciences. As a step towards a journal, she solicited short articles to be published in the newsletter. I still believe that such a periodical for southern Africa or perhaps for sub-Saharan Africa – or maybe even for Africa – would be a worthwhile goal to work towards in coming years.

WAY BACK IN 1959, when the very survival of the Wits Anatomy Department was problematical, many of these developments were merely dreams. From my experience of many societies and committees, I knew that dreams alone could generate little more than empty ideals and web-weaving. From the beginning, therefore, it was essential to take tangible steps forwards. In the Wits Surgery Department, my brother-professor, 'Sonny' du Plessis, had assumed office only six months before I did. He, too, had inherited a run-down, poorly endowed department and he, too, had high ideals. He was an exemplar for me in the Anatomy Department and, although I did not need any competitive goad to plan for a great and internationally reputable Anatomy Department, it was refreshing and stimulating to see how the Surgery Department, which had declined to a low ebb, was beginning a tide of resurgence that was to sweep it to an acme of excellence. It was indeed an admirable prototype of what we anatomists hoped to achieve – and, I believe, what we did achieve in the fullness of time.

So I, too, set about the tasks with vigour. I told my staff that I could obviously not do everything on my own. If the department was to flourish, I needed the devotion and help of every one of them. Through the periodic staff meetings, I would allocate duties to the team, but assured them that I would welcome their suggestions. At all times, they should feel free to discuss their and the department's problems with me: together we might plan the way forward. The response of Toby Arnold, Doris Bronks, Sarah Klempman, Ann Andrew, Mavis Feingold, Wendy Fine, Mary Veenstra, Hertha de Villiers, Adele Trope and 'table docs' such as Joe Borman, Myron Lange, Hiliard Hurwitz and others was heart-warming. Likewise the technical staff – Alun Hughes, Roland Klomfass, Sydney Dry, Dolly Spence, Ronnie Herman and Norman Harrington – rallied to the cause. The secretariat comprised Bernice Wilson for the first half-dozen years. As she had been Dart's secretary for some eight years, she was an invaluable link of continuity. My research assistants were Carole Orkin, Theo Badenhuizen, Jenni Soussi (later Tsafrir-Soussi), Adele Monro and Noreen Gruskin.

All of us dug and repaired and built. Quite soon the department was acquiring the semblance of a respectable structure. Funds came in from many sides. A range of researches and publications elicited more funds, more staff members, more research students – and so an increased research output.

It certainly was not roses, roses, all the way, and the apartheid juggernaut loomed ever more menacingly (see Chapters 26 and 27). Although peace of mind suffered smarting lapses and indeed went into abeyance for long periods, it was essential to keep going, to fight back, to lay more bricks in the erection of a department in which we and the university and post-apartheid South Africa could take pride. It was a kind of schizoid existence – upbuilding and rejoicing, alternating with tears of depression, elicited by doom-laden assaults on freedom.

In those days the appointment of a professor and head of department was to last for life, or at least until a fixed retirement age. In the first instance, therefore, I was appointed until the end of the year in which I turned 60. When that terminus drew near, the university extended my official tenure for another five years. So I remained head of the department until the end of 1990, the year in which I turned 65. What lay ahead of me then? The university in its wisdom decided to extend my appointment as a paid, full-time professor for a further three years, that is, until the end of the year in which I turned 68. The extension of a person's tenure to the end of the year in which he or she turned 68 was exceptionally rare at Wits at that time. I know of only two cases prior to mine and I felt honoured to have joined this small, select troupe.

During my lengthy term as head of department, the university had decided to go over to short-term appointments, of three or five years, to the headship. It would be wrong to call this system a 'rotating headship', for, at the end of a term of office a quite elaborate procedure would be set in motion, whereby the existing staff members of the department were consulted on their choice for the successor. The outcome could be for a re-appointment of the same person for a further term, up to a maximum of two consecutive terms, or some other existing member of the department might be selected. Under certain circumstances the university might advertise and, after consultations and interviews, appoint an outsider to the chair. The new system was thus based on short-term appointments. By the time my successor, Maciej Henneberg, came to be appointed, virtually every other department in the university had gone over to the new system. I suppose this was because of my inordinately long service. As long as I retained my sanity and sobriety, they could not kick me out until the compulsory retirement age prescribed in my contractual appointment was reached.

SPEAKING OF SANITY, it was not only staff members who might be or become unhinged. Students themselves, especially medical students, were liable to develop neuroses and psychoses and the university has an exacting duty in handling such 'cases'. In the Medical Faculty it is doubly difficult as, firstly, the demands of the course are especially exigent, and, secondly, students in this faculty are being trained to give service to the public. Thus a burden rests on the shoulders of the Medical Faculty to detect psychological problems early, before affected students are 'let loose on the public', as the unkind phrase has it.

I was so aware of the potential for psychological problems that I regularly entreated my staff, especially those who were themselves medically qualified, to watch out for early signs of undue stress. Moreover, each year I appointed to the part-time staff of the department a clinical neurologist, ostensibly to help usher the students into the wonders and intricacies of the nervous system, but principally to serve as a suitably qualified 'in-house' counsellor. By this mechanism many students with difficulties were detected and helped. Sometimes they would continue to pay visits to this counsellor even when they had moved on to senior years of study.

On some occasions I even used Jacobson's Progressive Relaxation Technique to help students who needed it. They would lie flat on their back, with a small book beneath their knees. Then I would instruct them to relax groups of muscles from the toes and feet, through the legs, thighs, abdomen and thorax, fingers, hands and wrists, forearms and arms and shoulders, neck and head – and lastly, and most difficult of all, to relax those fine muscles of facial expression, until the face was 'deadpan', expressionless. Initially it might take half an hour to achieve complete relaxation. Gradually over successive days you could achieve total relaxation in shorter and shorter times. Then totally relaxed, lie thus for about half an hour. Although the method was addressing the motor and proprioceptive systems of the body, it was amazing how relaxing an effect this exercise had on the mind. I taught many a student this technique.

Little is known, and nothing has been published, about this self-assumed role of the Anatomy Department and perhaps this is a good place to record it. I also practised what I preached. At times when I personally was under severe stress, I would restore my flagging energies by using Jacobson's system every evening when I arrived home. Then I would arise refreshed from a half-hour on the carpet, ready to embark on another night's work.

26 Apartheid in the university arena

For more than 40 years the universities of South Africa were under siege. The offensive started even before the general election of 1948 that unseated Smuts' United Party government and instated Malan's National Party government. Apartheid was a major plank in the nationalists' platform and the universities loomed large in their sabre-rattling. Threats at the hustings are one thing. Once the apartheid government was in power, the same and intensified threats became an ever-present sword of Damocles hanging over the heads of all who believed in freedom, equality and democracy.

Every part of society lay under the impendency of a loss of those cherished values. I chose to stay on in South Africa to fight against apartheid from within, at a time when very many of my fellow students and colleagues left the country. They departed for a variety of reasons: some because their political beliefs were under assault; a few as part of the 'normal' brain drain, moving to centres of excellence in other countries; others did not wish to take part in the compulsory military service that usually meant fighting on the frontier and killing black people; and some were lured by better economic opportunities elsewhere. Their departures left a great loneliness behind. Some of my closest friends from student days, those with whom I had grown and matured intellectually, were now far away.

It was, of course, my own choice to stay on in South Africa. I even declined a number of invitations to take up chairs from overseas. At the very moment when the chair of anatomy at Wits was placed in my hands in 1958, I received a dual offer of two chairs, one in anthropology and the other in human genetics, at the University of Michigan in Ann Arbor. I vividly recollect that their long-distance telephone call came through to me while I was attending a meeting of the Wits Senate. I had never received such a phone call before and the American dual overture brought on a tingle of excitement. It had to be weighed against the Wits invitation at the same time.

Michigan's offer was most attractive, and I have often wondered how different my life would have been if I had accepted it, or any of the other tempting proposals that came my way over the years from Canada, America, Britain, Israel, Australia and other anglophone parts of Africa. Nevertheless, I opted for Wits and assumed my position in the chair of anatomy on 1 January 1959. It was a fateful year.

I call it 'fateful' for it was the very year in which the apartheid government's long-threatened legislation to 'close' the 'Open' universities was due to come before parliament. For about eleven years our protests had been maintained. During that time I devoted a prodigious amount of time to speaking, organising, writing and demonstrating against the long-flaunted bill. On 22 May 1957, an enormous procession comprising the chancellor, council, Senate, lecturing staff, convocation and student body, some 2 000 people in all, marched solemnly and with dignity from the Wits campus in Braamfontein through the streets of Johannesburg, led by a single poster bearing the words, 'Against the Separate Universities Bill', to the steps of the Johannesburg City Hall. It was an orderly and quiet procession that received worldwide television coverage even before television was available in South Africa.

THE YEAR WAS fateful for other reasons, too. As you may have gathered by now, there were two sides to my make-up. On the one hand, there was my extrovert side: making speeches, organising and leading so many movements in the university and in the world of research. As an active head of a large university department, of necessity that was my public face, the face that my students, my staff and the public saw. On the other hand, there was my private side, sensitive, introspective and retiring, which the world did not see. When I was beset with melancholy and depression, I occasionally confided in my personal journal. I suppose that if I had ever had a wife it would have been to her that I would have turned at these difficult times.

Generally, I was too busily involved in writing, researching, lecturing and giving of myself to the department, the faculty, the Senate and Council of Wits, and the numbers of societies (local, national and international) of which I became an active adherent or of which I was the founder and nourisher. Overwork, I used to say, was the panacea for depression. Only on infrequent occasions over the last 50 years have I, almost in desperation, confided my thoughts and fears to my journal.

One of those rare occasions was in February 1959. It was exactly six weeks after I had assumed office as professor of anatomy. But it wasn't the challenges and burdens of running the Anatomy Department that were worrying me. Instead, it was the very problems about which I have been writing in this chapter. Forgive me if I quote at some length what my feelings were at that critical period when the university apartheid legislation was about to be laid before parliament:

> <u>Soliloquy on suicide</u>
> What has prompted me to [pick up my pen to write my thoughts]? Perhaps it is a profound and angry melancholy which has come upon me, at the impending destruction of the true nature of our university. I have just returned from discussions on plans to protest against the long-awaited university apartheid bill. I am filled with intense unhappiness, with a sense of frustration as I contemplate the 'closing' of our university. It is over ten years since first I launched the campaign against apartheid in the universities; and the sure knowledge that ultimate defeat looms near is a bitter prospect. For I no longer have any doubts that this legislation will go through parliament this year. Has the fight gone out of me? Have I so aged, or so weakened, that my old burning optimism no longer wells up in my bosom, overfloods my being, and bursts forth into serious action? Pray God this is not so. Perhaps the old energy is being canalised into new channels. Of course it must be, for it is true I am obsessed by the needs of my department. My plans for the department and my administration of it fill all my waking hours. I am obsessive. Yet it is not entirely true: for I weep with impotent rage at what lies ahead for academic freedom – and my thoughts are constantly flying on to the days when our university will be forced to deny its true nature, by becoming willy-nilly an apartheid institution closed to certain racial groups.
>
> The inspiration, the beatific experience, which the very living of a non-segregated existence in the university, has always meant to me, are so much part and parcel of my existence, of my conception of the

university, that I cannot bear the thought of the university without that essential ingredient.

Will [my] conscience allow [me] to continue on the staff of the university, to be a part of the system against which my very being cries out? Will [I] be able to tolerate it? Again and again, my mind flies back to the thought that I shall not be able to stay on, that I shall have to resign and depart. The problem is a tricky one . . .

Should I stay and fight on – in a stifling atmosphere – or would it make for more effective fighting to resign in protest? Is that sufficient of an act of fight, even though my own life may be ruined in the act? No, not 'even though' surely rather, <u>because</u> it would ruin my life, it would be the more glorious act of fighting!

I begin to feel – or perhaps I have felt for a long time – I wouldn't really care what happened to me after I had resigned. It's not that America is calling me, in spite of the fine offer I have had from Ann Arbor, Michigan. I do not think I would choose that as the alternative. What is the alternative? I do not know – my thoughts are a blur of ideas about serving Africa, about crusading for understanding and tolerance, about bringing people to think more rationally on matters of sex and human relations. Am I running away, or fighting the harder? . . .

In the last analysis, it has been said, most decisions in life are taken for selfish or personal motives. Ought I then to rate this aspect most highly? But, then, there is a reverse to this coin. On the personal side, it is not only a matter of conscience and of liberation: [I] stand to lose everything which has been built up over almost 15 years. [I] will be out of a job with very little money to my name. [I] will be unencumbered and somewhat impecunious. [I] will be faced with the possibilities of vegetating, of starving, of deteriorating, of mental collapse. In the face of such prospects, the realisation dawns that such a resignation will mean <u>intellectual suicide</u>. How big a step is there from intellectual to physical suicide? Would this resignation then be tantamount to committing suicide? And is my desire to resign not simply another way of expressing a suicide wish?

From the sheer misery reflected in my confidences to my journal on 11 February 1959, I bounced back. Seven weeks later, to a crowded Great Hall at Wits, I delivered an address on 2 April 1959. It was a meeting of the public of Johannesburg, convened by the Convocation of Wits University. I outlined the reasons why Wits and the University of Cape Town were united in their opposition to the implementation of the government's policy for the universities. I went on to point out that there were two sides to the coin of academic apartheid: the infringement of academic freedom on one hand, and the sheer racist thinking behind the impending legislation on the other:

> In the first place, the Bill constitutes a serious infringement of university autonomy and academic freedom. By telling the universities what classes of students they may admit and whom they may not, the state is arrogating to itself a right which the university has always previously exercised. The freedom to teach whomever it chooses is traditionally regarded as one of the four pillars of university freedom. As the Holloway Commission pointed out, the idea of academic freedom includes not only the freedom to search for the truth, but the freedom to impart the truth to others – and this implies, the freedom of others, such as students, to receive the information imparted. Any limitation of the right of certain types of students to enter the university thus does two things at the same time: it restricts the area in which the university exercises free and autonomous judgment; secondly, it mars the freedom of the university in its great tasks of seeking, conserving and imparting the truth.

I labour this matter of academic freedom and university autonomy because I felt then, and still feel very strongly, that a university must, by its very nature, be a world of freedom. A university, as a repository of learning, must remain free to study and discuss anything and everything. It must remain free to teach whomever it wishes to teach and to appoint as staff members whomever it wishes to appoint. Therefore, it must cherish and honour and maintain the very idea of freedom and, if necessary, speak out boldly in its defence.

In 1959 I went on to move the motion of protest, which read as follows:

> That this public meeting of the citizens of Johannesburg and the Reef notes with deep concern that in spite of the representations made to the

> Commission on the separate University Education Bill, the Extension of University Education Bill now before parliament embodies the same grave threats to university freedom and educational standards as were contained in the former Bill.
>
> This meeting affirms its faith in the ideal of the free and open university, and appeals to parliament to reject the Extension of University Education Bill *in toto*.

It was not long before our marches from Wits campus to the City Hall were banned. The students held peaceful demonstrations on the pavement along the east aspect of the university campus, facing on to one of the city's great arterial roads, Jan Smuts Avenue. Along this highway thousands of motor cars poured northwards, from downtown Johannesburg, every evening. Hundreds of students took part in these silent demonstrations, with posters that the occupants of passing cars could read. The student leaders invited me to stand with them in full academic dress – my scarlet doctoral gown faced with gold satin. Many passing motorists signified their support for our struggle by flashing their headlights as they passed us or sounding their hooters.

Unfortunately, not everyone was supportive of our cause. On several evenings, when the traffic thinned, a row of shouting thugs lined up on the opposite side of the road. As we were on the sidewalk of a public thoroughfare, we were sitting ducks for any projectiles hurled by them. The hoodlums took to the violent throwing of tomatoes, eggs and other objects at us. I took a hen's egg (unboiled) full on my red gown. For years I kept that mark, unwashed, on the gown as a macabre badge of honour. The men concerned were not in uniform and were very casually attired. We always assumed that they were students from RAU (the Rand Afrikaans University), a few kilometres west of the Wits main campus, but we had no proof of our suspicions – until one day about twenty years later! I was sitting with my carry-on baggage in the public departure area at Johannesburg International Airport. I saw a man sitting alongside me studying the label on my hand-baggage. He turned to me and said, 'Excuse me, but I note your name with great interest.' He went on to introduce himself and confessed that he and his fellow students from RAU were the ones who had lined up on the pavement outside the Lion's Brewery across the way from Wits. He confessed further that they were the chaps who had hurled missiles at us. Now at last there was corroboration that our opponents were from RAU. Then he made a further startling confession: 'D'you remember the eggs we threw across at the demonstrators?' I nodded and resisted narrowing my eyes slightly. 'Well, Professor, I was the one who threw the egg that hit your gown!' Slightly shamefacedly, he added, 'We were such hooligans then and we were naïve enough to like the government's policies and to despise what the Witsies stood for, but I've changed now – and well, I'm very sorry.'

NEEDLESS TO SAY, despite all our protests and much international support for our campaign, the bill went through parliament with a fearful clang of finality. The student leaders organised a symbolic dousing of the flame of academic freedom on the steps of the Central Block. At the students' request, I solemnly extinguished the flame. It was a moment of dewy eyes and barely suppressed embitterment.

I did not resign from the university. The fight went on for all of the years that the universities of South Africa were impaled on the stake of apartheid. After the eleven years of fulminating at the Open Universities from 1948 to 1959, there now followed twenty-five years of labouring under the apartheid policy, from 1959 to 1984. They were the Dark Ages of university education in South Africa. But the students and their courageous leaders, and the staff members, never allowed the matter to rest.

Plaques marking the closing of the Open Universities were erected and annually there was a commemorative address and a solemn reading of the wording on the plaque. The two universities most concerned, Cape Town and Wits, set up an annual freedom lecture sponsored by the chancellors of the two universities. Every possible route was explored by the university to get around the hated legislation that enjoined the need for ministerial consent for the admission of students of colour. As dean of the Medical Faculty in 1980–82, I found myself time and again pleading with Pretoria the cases of individual black students, on medical or compassionate or marital grounds! Indeed, we can proudly claim that throughout the twenty-five years of enforced apartheid, there was never a single year in which there was not at least one student of colour in the Wits Medical School. These were little more than drops in the ocean and the overwhelming majority of medical graduates from all the medical schools in South Africa continued to be white.

The problems at the level of admissions policy were only a part of the universities' difficulties. In Chapter 9 I mentioned our first on-campus encounter with a member of the security forces in 1949. Later, especially in the 1970s and 80s, there was a tightening of the noose around students who were considered to express their opposition to apartheid too vehemently. To this end, some students were surreptitiously recruited by the government's security *apparat* as informers. Others were planted in the student body and even rose to become high officers in the Students' Representative Council. As a result of 'information' gleaned by the Special Branch from such sources and from others, there was a spate of arrests and detentions. Needless to say, the victims were not brought before a court of justice.

By 1981 the intensification of the government's clamping down on all opposition had reached fever pitch. At a protest meeting at the medical school on 25 June 1981, which the students had invited me to address, I aligned myself with the concern of the medical students and expressed my solidarity with those who had been silenced. I made it clear that I was speaking in my personal capacity, not on behalf of the Faculty of Medicine (I was dean at the time), nor in my capacity as the senior representative of the Senate on the University Council.

In 1982, the third and last year of my term as dean, helped by Yusuf Dinath, the projects manager of the medical school, I analysed figures for medical graduates from all medical schools in South Africa. It was a shock to find that only 3 per cent of all of the doctors graduating in the country in the previous year were Black Africans. Yet Black Africans at that time comprised over 70 per cent of the total population. These appalling figures were given wide publicity in the media. Questions were asked in parliament. My statement from the dean's office at Wits was sent to the Ministers of Health and of Education and a great number of other authorities. Even the government (at last) felt that something must be done.

The 'solution' they came up with was based on racial quotas. If they dropped the highly discriminatory ministerial consent system, would the universities in question exercise a quota of non-white students? This proposal elicited widespread protests. On the Wits campus, the greatest mass meeting of members of the university took place. Speakers were drawn from all major subdivisions of the university. I spoke for the Senate and, to my amazement, received the biggest standing ovation that had ever befallen me. The quota idea quietly fell away. In 1984 almost all faculties were re-opened without any racial restrictions or quotas. The Wits Medical Faculty was excluded, obviously because the government wanted the brightest and the best of the African students to go to the apartheid showpiece, the Medical University of South Africa (Medunsa), a rural facility not far from the city of Pretoria, and intended for students of colour. In addition, for some reason that is still obscure, the course in Quantity Surveying at Wits was excluded from the newly regained freedom to admit students of colour without ministerial consent. We in the Medical Faculty did not take our continued subjugation to the ministerial consent clause lying down. Protests continued under the deanship of Clive Rosendorff and we brought those damning statistics up to date: they were still shockingly disproportionate. The overwhelming majority of doctors qualifying in South Africa came from the white minority, which comprised some 15 per cent of the total population at that time. Finally, in 1986, even as the State of Emergency pressed heavily on the supporters of the liberation struggle, all faculties and courses were opened to all students. It marked just about the only victory we had achieved in close on 40 years of dogged struggle.

27 Academic boycotts and freedoms

For a number of years during the apartheid era, scientists living in South Africa experienced difficulties in attending international scientific congresses in certain countries, including India, the Soviet Union and Japan. These difficulties usually resulted not from decisions by fellow scientists, but from the refusal of entry visas by the governments of the host countries. Such actions were intended as (to me) entirely understandable political gestures against the policies of South Africa, but at the same time they inevitably had the effect of infringing the freedom of movement of scientists.

This impact by the political arm on scientific freedom was not confined to South African scholars; at various times and places it was applied to Israeli scientists, scholars from the former Southern Rhodesia (now Zimbabwe), Soviet scientists wishing to attend congresses in Australia, Canada, and so on. To counter this infringement of scientific freedom, the International Council of Scientific Unions – the leading international co-ordinator of organised science – set up a Standing Committee on the Free Circulation of Scientists. Whenever one of the affiliated unions was subject to this sort of trouble, the Standing Committee sprang into action. In a number of instances it was able to press the host country to grant visas; now and again an international meeting was cancelled or the venue was moved elsewhere.

During the era of the academic boycott, several governments, such as that of the Netherlands, brought pressure to bear on universities in their country 'to minimise contact' with South Africa and South African scholars. One example of this kind of pressure was the case of a research student from Wits who had previously been invited to spend a few months in a Dutch university, to work on equipment and techniques needed for his doctoral studies. The invitation was withdrawn.

Another form of academic boycott was the refusal by the organisers to accept persons from South Africa at an international congress. The most widely publicised example was the so-called World Archaeological Congress at Southampton in England in 1986. The local organisers yielded to pressures on them to ban ('disinvite') the 26 scholars who were due to attend from South Africa. These included myself although I was a member of the Permanent Council of the International Union of Prehistoric and Protohistoric Sciences (IUPPS), one South African black archaeologist, several academics who had already been invited to participate in symposia at the Congress – and at least six British subjects. There were worldwide protests against the 'disinviting' of the South African delegation. Hundreds of intending participants cancelled their registrations. The presidents of the Royal Society (London) and the British Academy wrote a joint letter to *The Times* to express their objection to what the Southampton Organising Committee had done. The IUPPS under which the meeting was to be organised requested that the Southampton Committee reverse its decision, but to no avail. Consequently, the International Union withdrew recognition from the Southampton meeting and resolved to hold the IUPPS Congress elsewhere. The Southampton organisers went ahead with a 'rebel' meeting and although only 1 000 attended, instead of the 3 000 who had originally been expected, the Southampton meeting established a new international body, called the World Archaeological Congress. This has continued to assemble at intervals, independently of the IUPPS Congresses. Thus, the boycott in this particular case had unexpected consequences, namely the splitting of the international archaeological fraternity and the setting up of a new international congress series.

Other kinds of boycott arose. Not only were South African scholars debarred from an increasing number of international meetings, but some medical and other scientists from abroad refused invitations to attend congresses here. By attending such congresses, they argued, they would give credibility and support to the South African government's apartheid policies, 'or would somehow be tainted by their association with us'.

Another prong in this world attack was a tendency for an increased number of scientific articles submitted from South Africa to be rejected by editors of certain international scientific periodicals. There was even one example where a computer program in human genetics, which was freely available on the scientific world market, was embargoed as unavailable for sale in South Africa.

As the international academic boycott tightened its grip on South African universities, people like myself who had fought long and hard against apartheid and were anxiously awaiting the end of the despised regime, found themselves facing a dilemma. Should we support the academic boycott? Would it help to bring the Pretoria regime to its knees? I could foresee the part that the economic boycott would probably play, and the diplomatic boycott, and, in sports-mad South Africa, perhaps even the international sports boycott. The mass effects of these three kinds of boycott might well hit the South African population and the government hard and hasten the end of apartheid. But what of the academic boycott? Would it make any difference at all whether I was disinvited from a congress (as at Southampton), or whether a colleague was refused a visa to attend a scientific congress (as in Japan)? I venture to suggest that this kind of boycott would have had no impact whatsoever on the government and its policies. Their downfall would certainly not have been hastened by bans on a handful of South Africans from visiting laboratories (as happened in the Netherlands) or from publishing in an international scientific journal (as occurred in a periodical published in Berlin).

Unfortunately, while the academic boycott would undoubtedly leave the government in Pretoria unperturbed – if they even knew about the nuts and bolts of it – it was the universities of South Africa that I feared would suffer. Moreover, I prognosticated that there would be an increase in the brain drain of our brightest academics. It would be increasingly difficult for local academics to keep abreast of world developments in medicine and science. The standards of academics, of teaching and of research were likely to drop. Academic stagnation was the middle-term danger.

That was a prospect I could not contemplate with an easy mind. It would have been surprising if that was what our scientific and medical colleagues really wanted, but it seemed to me that it would be the inevitable consequence of the spread of the tactic of isolation. For that is what academic boycotts were – a tactic adopted by well-intentioned persons who saw with dismay that the fundamentals of apartheid had still not been eliminated. It was a tactic, I believed, that would backfire and destroy the great academic tradition that had been built up in South Africa.

Those who had grasped hold of the academic, cultural and medical boycott, as a tactic in the struggle, had sown the wind and we who were trying to keep the flame of higher education alight in this country would reap the whirlwind. I understood their tactic; I supported the purpose – to end apartheid – but if the international cultural boycott succeeded, its proponents would have dug the graves of this country's universities.

There was no chance that an academic boycott would give a boost to local universities – for universities do not flourish in isolation. One example is furnished by the history of genetics in the Soviet Union. During the era of Lysenko, the Soviet Genetics Decree of 1948 cut off 'Soviet genetics' from genetics proper, which was built on the work of the monk, Mendel. Ultimately, after years in a genetic wilderness, the Soviets saw the immense success of Western genetics – the double-helix structure of DNA was discovered, the genetic code was cracked, genetic engineering was launched.

Every biology textbook in the Soviet Union had to be destroyed; every book had to be rewritten with the new genetical knowledge; all teaching of biology had to be stopped for a year. Universities need the cross-fertilisation of international contacts, not only to flourish, but even just to keep up to date.

Of course, but sadly, the warnings and reservations that I gave voice to cost me some respect among the external liberation movements.

Happily, ten years after the birth of democratic South Africa, such boycotts are a thing of the past. Now the policy of the ICSU (International Council of Scientific Unions) prevails generally – that in matters of scientific exchanges, invitations, representations and publications, there shall be no discrimination on grounds of race, colour, creed, nationality, political persuasion, gender, language or sexual preference. Moreover, South Africa has become a popular venue for international meetings and fine international conference centres have arisen in Durban, Johannesburg and Cape Town.

A number of my overseas scientist-friends would not on principle come to South Africa during the apartheid regime. Their attitude was thoroughly understandable; of that I was never in any doubt. It was their laudable gesture against apartheid. Yet it was desolating to those of us who had chosen to stay and to fight it out from within. Fortunately, I have lived long enough to witness and feel the joyous arrival of the New South Africa and to find myself operating in a normalising academic community. My friends far away are now happy to visit us and I welcome them back with open arms.

I WAS FIRST inspired to put forward the Four Freedoms of the Academy by the graduation address delivered to Wits in the Johannesburg City Hall on 16 March 1946. I have already touched on that occasion (in Chapter 9) when Jan H. Hofmeyr, the chancellor of Wits, reminded us that the Four Freedoms of the Atlantic Charter should serve as the measuring rod in the continuing battle for freedom within South Africa. Those Four Freedoms – the freedom of speech and expression, freedom of religious worship, the freedom from want and the freedom from fear – and Hofmeyr's proposal that a Fifth Freedom – the freedom from prejudice – be added to the list formed the backdrop to the Four Freedoms of the university world that I enunciated in July 1949, just over three years later.

We in the world of the university have long recognised our own Four Freedoms. They are the freedoms that we believe are essential to the proper functioning of the university and to the flourishing of the human intellect. They are components of what is often called academic freedom. In July 1949, one year after I had become president of Nusas, the National Union celebrated 25 years since its establishment, with a Silver Jubilee Congress held at the University of Cape Town. At that meeting I enunciated, I believe for the first time, the Four Freedoms of the academic world. In the presidential address, I stated:

> . . . Nusas must maintain a constant vigil over the right of the universities to determine who shall become students; the right to determine in what manner classes will be organised within the university, including classes for groups of students of varying racial origin; the right to appoint members of the academic staff without reference to governmental authorities; and the right to conduct its courses as it sees fit, and not in accordance with any state-imposed system of thought and philosophy.
>
> These rights are of the very essence of university freedom, for truth in a strait-jacket is not truth; the university in political chains is no university.

This first formulation of the Four Freedoms, with which I received help from Ieuan Glyn Thomas, the registrar and later vice-principal of Wits, was adopted by Nusas as its official policy and in short succession was also taken up by the highest academic authorities in the country. When representatives of Wits and Cape Town Universities came together in Cape Town on 9, 10 and 11 January 1957, to prepare the now famous little work, *The Open Universities in South Africa*, these rights were enshrined in that book as 'the four essential freedoms' of a university.

EVER SINCE FIRST enunciating them, it had worried me that the Four Freedoms emphasised only the freedom of the 'academy' and of the senior academics, that is, its staff. Even the freedom to teach whomever it wishes meant a freedom of the university, not a freedom of the junior academics, that is, the students. Indeed, initially it seems to have been orthodox at least in the English-speaking world, to deny that the idea of academic freedom had any application to students.

With this in mind, I proposed that a Fifth Freedom be added to the list. This Fifth Freedom was related not so much to the academy or university as to those who were freely able to attend the university. It steered emphasis away from the university itself – with which the Four Freedoms dealt – on to the students or potential students. The establishment of ethnic or tribal universities in South Africa and the 1959 legislation, euphemistically dubbed 'The Extension of University Education Act', brought about a shocking form of unequivocal discrimination against black students. White students were free to attend any university they wished to, provided they were academically eligible and were selected (in the case of restricted faculties). They had an array of choices among a set of fine, long-established institutions, some with international reputations. Black students, by contrast, were allowed to attend only the university or college that pertained to their ethnic group. They might apply to what were called in apartheid South Africa, 'White Universities'; but, in terms of the legislation, they might not be admitted to these universities without the prior permission, in each instance, of the ethnically relevant Minister.

The position was particularly serious in the case of medical education as the government was attempting to confine black medical education to a single African 'Medical University' at Garankuwa near Pretoria to serve the country's entire African population. This denial of the freedom of choice to blacks who wished to become medical students was a blatant form of discrimination against African students that could never be offset by the trappings, however fine, of one institution near Pretoria.

The Fifth Freedom, which should be considered part of academic freedom in the broad sense, was formulated as follows: all students or potential students should have the freedom to select the university at which they wish to study and, if they satisfy the academic requirements and are selected, be free to attend such university without governmental interference and without the need to seek special permission from any other authority. This freedom of the student to choose his or her university complements the freedom of the university to choose its students without let or hindrance.

In time, I came to propose a Sixth Freedom that also pertained to students and their rights in the university context. It was enunciated as follows: students should have the right to organise and control their own affairs, free from governmental and outside restrictions; they should have freedom of assembly, freedom to hold what beliefs they will and freedom to express their opinions with impunity. Closely coupled with this freedom is the right of students to participate in the governing bodies of our universities.

This Sixth Freedom was really not a new idea. It was proposed as part of a students' credo, an epitome of the spirit of Nusas, in that same presidential address delivered by me to the Silver Jubilee Congress of the National Union in July 1949.

I later formally proposed that these rights and liberties be recognised and enshrined as a Sixth Freedom in the Edgar Brookes Academic and Human Freedom Lecture of 1976, at the invitation of the Students' Representative Council of the University of Natal, Pietermaritzburg, on 6 May 1976.

TIME DOES NOT allow me to detail the history of outside political pressures and outright attacks on Nusas from the late 1950s to the 70s and 80s. I should like, however, to refer to one of the attempts of the government to clip the wings of Nusas. There are more ways of killing a cat than choking her with cream! Consequently, the government, while at first shrinking from outright banning of Nusas, tried year after year to kill Nusas by cutting down its leaders. This failed to break the National Union – through the sheer courage of brave new young men and women who continued to come forward as national student leaders, despite the dangers and penalties to which they were exposed. Then, the government resorted to starving the National Union to death by cutting off some of its finances. On 13 September 1974, the Minister of Justice announced in the House of Assembly that the government had declared Nusas an 'affected organisation' and a notice to this effect was published in the Government Gazette shortly after the announcement. This effectively prohibited Nusas from receiving donations from abroad.

The government chose to lay before parliament and the public of South Africa an entirely one-sided view of Nusas, ignoring all the good work that it had done over many, many years, much of it with overseas funds. It damned the student organisation, largely because of the alleged political activities of some of its leaders, without any charges being tested in a court of law. I believe that the government overreacted to reports and enquiries into Nusas, while the students themselves were not given the opportunity of having their side of the case heard and evaluated by an objective public judicial enquiry.

There are national student unions in virtually every country in the world that boasts universities. All of them – and Nusas was no exception – have a great number of practical and socially useful projects, like the African Medical Scholarships Trust Fund (see Chapter 9). All of them, too, express the idealism and the radicalism of intelligent young students. In like vein, Nusas was a power of good in practical projects for student welfare and for the community. Further, it stood out as one of the most vocal forces in the country holding out against apartheid and believing in the ideal of a non-racial and common society, free of discrimination and segregation. It was the power of Nusas' opposition to racialism and to the racial basis of our South African society that, I believe, produced the apartheid government's 40-year-long vendetta against the National Union and its decision to try to paralyse Nusas by declaring it an affected organisation.

I deplored this most retrogressive step and urged the government to have second thoughts, commenting in a speech at a protest meeting in the Wits Great Hall:

> Leave the students to run their own affairs as they wish, within the letter of the law. It would be sad indeed if the public of South Africa and the student body were to be misled by the one-sidedness of the procedures the government has adopted, to crush the spirit and virtually to deny the right of existence of Nusas.

For this body was in the vanguard of advanced constructive thinking about South African society for over 70 years and about academic and human freedom for some 50 of those years.

A country in which the government denies to some parts of its population the Fifth Freedom, freedom of choice of a university, and to some of its students, the Sixth Freedom, freedom of organisation and assembly, is not only impairing academic freedom; it is impoverishing itself in matters of the mind and spirit. For freedom is essential for the supreme blossoming of the human intellect, for the preservation of humankind's creative genius, spontaneity and originality, as sustaining forces of our universal cultural legacy. Freedom of organisation by students is an essential part of that cultural legacy.

In those dark days, the 40 lean years, I drew consolation and inspiration on many occasions from Carl Sandburg's lines in his 1936 poem 'The People Will Live On':

> Man is a long time coming,
> Man will yet win.
> Brother may yet line up with brother.
> This old anvil laughs at many broken hammers.
> There are men who can't be bought.

And, indeed, the passage of time thankfully has proved just how correct Sandburg's words were.

28 Joyous variety

Each year, welcoming new classes to their debut in anatomy, I shared Samuel Johnson's sentiment, 'The joy of life is variety' (*The Idler*), when I contemplated the pleasing array of several hundred students whose eager faces in a crowded auditorium were fixed on mine. Whilst delivering the welcoming oration, I had time to study the infinite variety of their visages. They were long-faced and squat-faced, blue-eyed and black-eyed – and joy of joy, some with one blue eye and one brown eye (known to the cognoscenti as *binocular heterochromia iridis*). Some smiled, others scowled; a few lips quivered tremulously; some had narrow foreheads and some broad foreheads – so very broad in fact as to lead me to infer that we might be sitting in the presence of metopism, that is, a cranium in which the suture between the left and right frontal bones had not fused in earlier life, so that growth in the breadth of the forehead went on beyond the usual limit. Some had perfectly formed features; some had surgically repaired cleft lips. There were narrow, high-bridged noses and broad, low-bridged noses; high, prominent cheek-bones or low, rounded cheek-bones; pallid and tanned white skins, or light brown or very dark brown skins; unornamented or bejewelled or tattooed noses, brows, eyelids and ears, nature's rich diversity supplemented by cultural variegation; some with rich hues imprinted by recent summer vacations and some shielded, blocked out, parasoled and veiled. In the panoply of what Archbishop Desmond Tutu called the Rainbow Nation, some wore religious insignia, the imprints of political affiliations, sexual signposts and such marks of male development as beardlessness or facial fluff or prickled, sprouting beard and moustache. What an exercise in human biology and variation was laid out there before me, year after year.

When I was a lecturer and senior lecturer under Raymond Dart, I and others at that stage were permitted to invigilate in the university's main examination halls. This gave me a fantastic opportunity to count the numbers of left- and right-handed students. Moreover, I could compare the rows of engineering students with those of arts and of health sciences students. To my amazement, I found that among medical students, left-handedness was twice as frequent as among the engineers and arts students whose data matched those for the general population. Year after year, I found this same result. By chance, a good friend of mine, Dr M.D.W. ('Doc') Jeffreys who was a social anthropologist, found precisely the same correlation. For some reason it was more likely to find left-handers drawn to medicine than to other disciplines. As to dentistry and physiotherapy, the figures of those classes also were elevated above those of the population at large, but not as markedly as among the medical students.

Was there something about the healing sciences that drew to their fold people who were perhaps a little different from the common herd?

Doc Jeffreys and I often spoke about preparing a joint paper on the incidence of handedness; alas, we did not get around to it – and this memoir affords me the first opportunity to set down these thoughts. Perhaps someone will pick up the challenge, although of course methods of student selection these days are very different from what they were 50 years ago.

IF FACES AND writing hands reveal much about the fantastic amount of variety among males and females of the human species, just imagine the miscellany of physiques that are laid bare in unclad, or even partially clad, subjects. Why anyone should want to observe and compare human physiques is a question that needs an answer – and physicians, psychiatrists, Renaissance artists and Olympic athletes may well give us some clues. Before I turn to these beguiling questions, let me say a few words about the methods used in human biology to characterise the varieties of human physique (this last phrase was the title of William H. Sheldon's well-known book of 1940). It is a cruel and unscientific practice of boys and girls, and men and women, to call a person 'fatty' or 'skinny' and then to assign all manner of attributes to them, such as chubby and lovable, or serious-minded and unfriendly. Success in the swimming pool, or the sports field, or sexual hunger and prowess – are among the many features that are, often unthinkingly and unkindly, attributed to extreme examples of the kinds of humans. Novelists, too, are especially inclined to use such phrases as sensitive fingers, sensuous lips, a world-weary brow, and so on.

How much of this is accurate and how much simply vague impressions? Several techniques are available for the study of physique. The most widely used up to the third quarter of the 1900s was one or other modification of William Sheldon's method, based on standard photographs – from the front, the side and the rear – of each subject. Apart from the photographic technique, there are several methods based not on photographs but on simple inspection and scoring of the parts of the body. I used a simplified inspectional technique in my field studies of the Kalahari San (see Chapter 10) and the Zambian Tonga (see Chapter 23).

In June 1956, I paid an interesting and revealing visit to William Sheldon himself and to his somatotype studio in Columbia University, New York City. He was passionate and obsessional when it came to classifying the bodies and minds of his students and others about him.

Sheldon's method differed from those of physicians and psychiatrists before him, in that from his study of 46 000 subjects, he was able to recognise three components or physical tendencies in living humankind. When the *ectomorphic component* predominated, the subject was thin, light-boned, linear, flat-chested, with a small trunk and recessed abdomen, and relatively long limbs. When *mesomorphy* was pre-eminent, the individual was broad, big-chested, heavy-muscled, large-boned and big-hearted. When the *endomorphic component* preponderated, the body was broad and rounded, inclined to fat, big in the stomach, light on muscle, approximating a sphere with short limbs. Everyone was considered to be a mixture, in varying degrees, of these three components. The *somatotype* was the formula or personal recipe that expressed the relative proportions of the three components in each individual examined.

One drawback of the Sheldonian form of analysis was that he used a closed-ended scale (1 to 5). It was this method that Wesley Dupertuis of Cleveland was planning to apply to the Kalahari San, while I used a modified inspectional method. However, Barbara Honeyman-Heath of Carmel, California, and Lindsay Carter of San Diego, California, had found this use of a closed-ended scale most unsatisfactory. For instance, when Heath had worked with Margaret Mead on Manus Island in Melanesia, she found powerfully built mesomorphic men who so far exceeded Sheldon's maximum mesomorphic values for American students that, instead of his maximum of 5, they would have required a score of 10 to 15! So Heath and Carter introduced an open-ended scale, which, it was hoped, would be able to embrace even the most extreme variants of living humanity.

Two colleagues, Roy Morris and Leopold Jacobs, in the Department of Medicine at Wits University, devised a method of scoring somatotype formulae without resort to nude photographs – always a tricky problem. As in Sheldon's method, the body was divided into five regions. For each region, seven features were identified and each was scored as the endo-, meso- or ectomorphic manifestation of that feature. This inspectional approach had the great advantage that it could be used in the field. I first applied their inspectional approach in my field studies of the Kalahari Bushmen and, in 1957–58, of the Tonga people living in the Gwembe Valley of Zambia's Southern Province. I was able to present the first paper on the somatotypes of black people of southern Africa for publication in the *Journal of Human Evolution* in 1972. I found that the African subjects from Zambia were high in mesomorphy and ectomorphy but largely lacked the endomorphic component. Phyllis Danby of Oxford found an almost identical pattern in her somatotype study of East African black people, the only other study that had been made up to that time on the physiques of black Africans.

CLEARLY THERE ARE varieties of physique and for centuries people have wondered what they meant. In the world of sport, good swimmers need a fair amount of fat – it helps them to float more easily – and strong muscles for propulsion through the water. High jumpers require long lower limbs. The game of rugby needs light, agile, swift ecto-mesomorphs in the three-quarter line, and massive, meso-endomorphic Angus steers in the scrum!

Sheldon's big book, *Atlas of Men* (1954), is a remarkable work, deeply perceptive and, in parts, hilarious as he intimates the personalities and animal equivalents of each somatotype. For example:

> Somatotype 126: *on the mesomorphic side of extreme ectomorphy: 'Little wasps. Slight, delicate fellows, crushed by your lightest step. Yet they can sting.'*

Somatotype 145: *meso-ectomorphs, with no detectable endomorphy: 'Lesser falcons. Sparrow hawks. Merlins. Fragile, graceful, but incredibly agile hunters, who are rapacious eaters of fresh meat.'*

Somatotype 424: *Far from the mesomorphic north pole, with balanced doses of endomorphy and ectomorphy: 'Oppossums. Delicate, furtive marsupials who hunt beetles at night, or whatever such scraps they can find, and like the bears and woodchucks get fat for the winter. Nearly innocent of mesomorphy, their most effective defense against attack is to "play possum".'*

William Sheldon kindly gave me a copy of *The Atlas* and inscribed it to me in red crayon. Hesitantly I enquired, 'What about female somatotypes?' *The Atlas of Women*, he told me, was nearly finished and he hoped it would appear soon. He was also envisaging an Atlas of Children.

OVER 2 000 YEARS ago, Hippocrates, the Father of Medicine, classified humankind into varieties of physique and ascribed particular weaknesses to each. Long, thin, vertical bodies, he said, were more likely to develop chest trouble and phthisis (tuberculosis). Short, thick, horizontal people had a greater tendency to diseases of the heart and blood-vessels.

Later, psychological attributes came to be associated with physique. In art and literature such usages are common and often show the most extraordinary insight. I am thinking, for instance, of the physiques and personalities of Mephistopheles (shown always as a lean, intensely ectomorphic figure), Falstaff (broad, jovial and endomorphic), the endomorphic Buddha and ectomorphic Jesus, the comedy pair of Laurel (ectomorphic) and Hardy (endomorphic or endo-mesomorphic).

The clearest statement of all came from Shakespeare's *Julius Caesar*:

Caesar: 'Let me have men about me that are fat;
Sleek-headed men and such as sleep o'nights;
Yond Cassius has a lean and hungry look;
He thinks too much: such men are dangerous.'

Anthony: 'Fear him not, Caesar; he's not dangerous;
He is a noble Roman, and well given.'

Caesar: 'Would he were fatter! – but I fear him not:
Yet if my name were liable to fear,
I do not know the man I should avoid
So soon as that spare Cassius. He reads much;
He is a great observer and he looks
Quite through the deeds of men; he loves no plays,
As thou dost, Anthony; he hears no music;
Seldom he smiles, and smiles in such a sort
As if he mock'd himself and scorn'd his spirit
That could be moved to smile at anything.
Such men as he be never at heart's ease
Whiles they behold a greater than themselves,
And therefore are they very dangerous.'

Even among doctors there appears to be a preferred type. It is a common observation that surgeons are more mesomorphic than physicians, psychiatrists more ectomorphic than obstetricians – whilst anatomists (I like to think) are good, average, well-mixed, middle-of-the-road chaps!

SHELDON'S EFFORT TO correlate physical appearance with personality and behaviour had a long history. Many attempts to classify human physique had been made in the nineteenth century. Ernst Kretschmer, in particular, had tried to correlate the broad, bulky individuals with a cyclothymic personality and, in pathological cases, with the circular or manic-depressive psychosis; and the long, 'lean and hungry' individuals, in pathological manifestations, to schizophrenia.

The principal drawback about Kretschmer's approach and those of other early classifiers of humanity was that they divided people into 'types'. It is said that Kretschmer came on his ideas when he noticed striking differences in the physiques of his patients, when he walked across from the wards for manic-depressive psychotics to those for schizophrenics: the former tended to be thick-set, muscular, broad-built, while the latter were slender. This correlation was not established by what we would regard today as good statistical means: it was merely a tendency. If we took a sample of subjects and placed each into one of Kretschmer's original set of types, only just over a quarter of all people could be classified into one or other of these types. What of the remaining people? It was clearly not a very suitable way of examining physique and temperament. This lingering uncertainty led Sheldon to abandon the *types* of individuals and to recognise instead the *components* of physique. This approach recognised that each person had his or her own private recipe comprising proportions of each of the three ingredients.

When this system worked for somatic features, Sheldon sought to establish a similar way of looking at personality and behaviour. He found three psycho-components. The first component was called *viscerotonia* and if you have a large dose of it, you are likely to show relaxation, extraversion of affect, to love food and sociality. The second component of behaviours, *somatotonia*, is marked by bodily assertiveness and a desire for muscular activity. People showing the third component, *cerebrotonia*, are liable to be hyperattentional and to inhibit the other two components. When Sheldon examined the somatotype and psychotype scores, he claimed to find a very strong correlation between them. For instance, he held that a strongly mesomorphic person was likely to be markedly somatotonic. One serious criticism was that the same person – Sheldon himself – carried out both the somatotype and psychotype studies. Can we be absolutely confident that bias did not enter these analyses? It is surmised that Sheldon would gaze at his students and his interlocutors with an intensely scrutinising look and you could almost imagine him saying to himself, 'Hm, a touch of this I detect', and 'Aha, a smidgin of that'. Sitting opposite him at lunch in the canteen of the College of Physicians and Surgeons at Columbia University, New York City, was a discomforting experience. I found myself looking back with an equally scrutinising, staring gaze, thinking to myself, 'Ah yes, a pinch of this' and 'Begorra, a soupçon of that', but obviously with far less nous than the maestro himself!

The study of physique was a worthwhile undertaking, but I feel its marriage to these temperamental, behavioural and psychiatric aspects is one of the critical factors that led this branch of physical anthropology to fall into disrepute among many human biologists. Another factor conducive to its decline was the tendency among some of the biotypologists, especially the earlier ones, to equate the psychotypes with various racial groups. This was one of the underpinnings of the racialistic anthropology of some Nazi biotypologists such as H.F.K. Günther. Another aberration of such studies was the attempt to associate physique and criminality, to which Hooton's researches mentioned earlier gave the quietus. For these socio-political reasons, the study of physique became unpopular in some anthropological quarters. However, in medical circles and in sports medicine and physiology, somatotype studies are still considered useful. Two of my Ph.D. graduates devoted their theses to sports physiques, one (Bruce Copley) to tennis players and one (Paul Smit) to motor activities of South African children of varying nutritional status.

AS ALWAYS, I have enjoyed a little evolutionary exercise on somatotypes. In modern humans, variability is the spice of life. Population by population, we find percentages of the various somatotypic components. What has exercised my mind is this: in our early fossil hominid ancestors, were the three components present and, if so, were they represented in roughly the same proportions as in modern humanity? It is difficult to find an answer, as we are dependent exclusively on the bones and joint surfaces. But the bones are tell-tale. The bones of heavily muscled mesomorphs bear powerful muscle-markings. Those of mainly ectomorphic and endomorphic subjects may be expected to be more slender, more lissom. On this criterion I was once led to conclude that our ancestors were more mesomorphic and that the lithe bones of those in whom the other two somatotype components were predominant were less frequently encountered.

Although I have not published on this subject (as far as I remember), I have in lectures to my students and to the public now and again suggested that the early hominids were essentially the rough-and-ready mesomorphs. As their bodily structure modernised, a process of *gracilisation* developed. In other words, a bigger percentage of the population began to show endomorphy or ectomorphy or both. I am sure Sheldon would have exclaimed, 'Of course, for those are the two somatotypes that go with the awakening and flowering of culture – the ectomorphs being the thoughtful, brainy chaps who invented and promoted culture; the endomorphs with their conviviality, their sociality, providing the social milieu among whom culture is likely to have been nudged ahead and nourished.' But Sheldon would have been guilty of a circular argument by resorting to the behavioural counterparts of different somatotypes. I have never developed the theme along those lines, but had I had the chance in later life to put the idea to Sheldon, perhaps another of his poetic and insightful – and misleading – books might have appeared.

29 | My most unforgettable character

The Reader's Digest used to have an occasional feature called, 'My most unforgettable character'. Without any doubt, Raymond Dart would qualify as *my* most unforgettable character. He was head of the Wits Anatomy Department from 1923 to 1958. For eighteen years of that time, he was also dean of the Faculty of Medicine. His deanship started in 1925, the historic year in which he presented to the world his revolutionary interpretation of the Taung skull that had been found in 1924.

The year 1925 was also when I was born. With some small smattering of the gestation period in humans, I worked out that, if my birth on 14 October 1925 followed an average duration of my mother's second pregnancy, I would have been conceived on or about 3 February, the very evening on which *The Star* of Johannesburg carried the first exciting announcement of the discovery of the Taung skull. I have no reason to suppose that this world-shaking event (the Taung child's 'birth') influenced my parents' nocturnal behaviour patterns in any way, but it is a coincidence that I enjoy taking note of, in a whimsical mood of historical continuity.

Dart's deanship came to an end in 1943, the year in which I entered the Faculty of Medicine as a first-year medical student. Thus it came about that when my class was ushered into the faculty at the opening of the first year of our six-year course, it was not Dart who welcomed us with one of his customary orations, but his successor as dean, Professor John Mitchell Watt, the head of the Department of Pharmacology. I recall Professor Watt arriving on the scene complete with a cravat, morning jacket and striped trousers, accompanied by a lackey – in the person of one of his technical staff in a suit. It was a pompous occasion of unduly grand style. We were all bowled over, not to mention suitably impressed.

It was only a few weeks later that we had our first encounter with Raymond Dart. As he had served for eighteen years as dean, the Students'

Medical Council (SMC) organised a farewell presentation to Dart. The entire medical student body was summoned to the Great Hall of Wits University. The president of the SMC delivered a superb homily and expressed the gratitude of the students to their outgoing dean. As a gift of thanks he presented Dart with a set of *Encyclopaedia Britannica*. Dart tried to give voice to his appreciation. Until then I had never seen a grown man weep in public. It was a profoundly moving moment for all of us and it was my first introduction to the deeply emotional side of the man.

I had another experience of it a few years later. Robert Broom's 80th birthday took place in 1946. As chairperson of the Science Students' Society (converted to the Science Students' Council that year), I persuaded the Council to organise a tea party in honour of Dr Broom and to make a presentation to him. He was, after all, one of South Africa's most distinguished scientists, a Fellow of the Royal Society of London, and our teacher in human palaeontology in the Anatomy Department. I had invited Dart to give the main speech. He was sitting alongside me at the head table and broke down completely for some minutes early in his address. Mrs Dart's reassuring hand-squeeze helped to restore his self-control. He apologised to the audience, attributing his tears to the close links that had existed between the two scientists during all those difficult years of standing alone against a hostile world.

FIFTY YEARS PRIOR to his retirement from the deanship, Dart had been born in Toowong, now a suburb of Brisbane, Queensland, Australia. He entered the newly opened University of Queensland in 1911, graduating with a Bachelor of Science degree in 1913. He moved to the University of Sydney where he studied medicine. His mentor in anatomy was a powerful and persuasive figure, James T. ('Jummy') Wilson who later took the chair of anatomy at Cambridge. Dart completed his Master of Science degree in anatomy and his medical degree in 1917. Then he saw war service in the Australian Army Medical Corps in France and England in 1918–19. When he was demobilised he was invited to join the staff of the Anatomy Department under his fellow Australian, Grafton Elliot Smith, at University College, London. Dart's verve and intensity were already apparent. One night, when he needed to pursue his research after hours, he broke in to the

Anatomy Department and, so the story goes, was forthwith arrested. That's enthusiasm for you!

In the early 1920s, the Rockefeller Foundation of the USA desired to set up fellowships for foreign graduates, especially from the British Commonwealth, to the United States. The scheme came into being in 1920 when the first two fellowships were awarded to the young Australians Raymond Dart and Joseph Shellshear. During his time in the USA, Dart worked for a while at the University of Cincinnati. There he met an anatomy tutor and medical student, Dora Tyree, whom he wedded. Later, as Dora Dart, she endowed a chair of medicine at Wits.

Dart spent most of his time in the USA working at Washington University, St Louis, Missouri, in the Anatomy Department under Robert James Terry. By a strange turn of the wheel, 36 years later when I became a Rockefeller Fellow, I, too, spent valuable time at Wash U., where octogenarian Terry still paid a daily visit to the Anatomy Department. He took a keen interest in my undertakings there, as he had done with Dart's so many years previously. Once he came into my lab, patted me on the back, and said, 'Well young Raymond, how are you getting on?'

When Dart arrived back from America he spent a further short spell with Elliot Smith at University College, London. Then, rather against his will and under pressure from Smith, Dart accepted appointment as the professor of anatomy at Wits, with effect from the beginning of 1923.

Dart succeeded Edward Philip Stibbe, the first incumbent who had a short-lived (and controversial) stay at Wits. In all the years in which I was closely associated with Dart, I did not once hear him mention the name of Stibbe. To all intents and purposes, until Paton's book *Hofmeyr* appeared in 1964, it was as though there had never been a professor of anatomy at Wits before Dart's accession in 1923. In fact, students and younger staff members assumed that Dart had been the first professor of anatomy at Wits.

When I became head of the Department of Anatomy I realised that there was no single portrait or mention of the name Stibbe in any part of the Anatomy Department. I was determined to set this right. Finally, a measure of belated justice was accorded to Stibbe when an enlarged portrait of him was hung in the department's Hunterian Museum; and, further, when the postgraduate lab was renamed the Edward Philip Stibbe Postgraduate

Laboratory. It was most appropriate that the man who first presided over our Anatomy Department and was responsible for the first teaching of human morphology and for the first dissection of the cadaver between Cape Town and Cairo should have his pioneering role suitably recognised.

How quickly the past can be forgotten if we do not acknowledge our debt to the pioneers who came before us and on whose efforts our own endeavours are frequently built.

DART WAS A man of extraordinary vision. He came to a fledgling medical school, which had not yet graduated its first class of medical doctors. There was not even a medical library yet. Blessed with foresight, energy and physical strength, Dart played a major part in building up the school and its courses for medical and other health science students. As dean, he was responsible for bringing the Witwatersrand Medical Library into being; for setting up a degree course in physiotherapy (physical therapy); for cementing the relationships between Wits and such bodies as the South African Medical and Dental Council, the South African Nursing Council, the Medical Association of South Africa, the Wits Medical Graduates Association – and more besides.

It would be no exaggeration to say that Dart was the chief architect of the Wits Medical School until his retirement from the deanship in 1943. At the same time, he built up his own department. An early venture was the initiation, jointly with the Faculty of Science, of a Medical B.Sc. degree course. This grew to be a much-valued asset of the school and was something that I myself pursued with great enthusiasm (see Chapter 5).

In the first year of the new Medical B.Sc. course, one of Dart's students was Josephine Salmons. In 1924 she brought to her professor a fossil baboon skull embedded in breccia that had been collected by E.G. Izod at the Buxton Limeworks at Taung (at that time known as Taungs). That specimen excited Dart's interest. It was to prove a link in a chain of occurrences leading to the recovery, later in the same year, of the famous little Taung skull. Miss Salmons had played a crucial part in the unravelling of Africa's first 'missing link'.

The Taung skull was found in a limestone deposit at the Buxton Limeworks near Taung in the northern Cape province (as the area was known at the time). Having chipped the fossil out from the solid breccia in which it was entombed, in the weeks between the discovery date in November 1924 and Christmas 1924, Dart descried an unprecedented *mélange* of traits. The young professor pondered over this pot-pourri of unexpectedly concatenated features. What game was the soil of South Africa playing on him? Of a trice he remembered and reread the long-forgotten words of Charles Darwin. In his second great book *The Descent of Man* (1871), Darwin had written that 'it is somewhat more probable that our early progenitors lived on the African continent than elsewhere'. Raymond Dart made an insightful intellectual leap: he declared that the Taung skull provided evidence of the former presence in Africa of an anthropoid ape that, in some features of its morphology, had moved substantially in the human direction. He made this claim in an article in *Nature* on 7 February 1925. In brief, Dart recognised it as a higher primate that, though basically an ape in Dart's thinking, belonged to a form that had taken decided steps on the road to humanity. He named it *Australopithecus africanus* – or the southern ape of Africa.

It was clear to Dart that *Australopithecus* did not fit within what was then called the ape family or the Pongidae, but he could not bring himself to place this small-brained creature within the family of humankind, Hominidae, as it was then recognised. The brain size was smaller by far than those of hominids; it was scarcely larger than those of apes. Instead, he proposed to create for it a new, intermediate family, which he suggested should be called *Homo-Simiadae* ('Man-Ape Family').

However, with very few exceptions, this suggestion was not supported. From early on, Robert Broom backed Dart's claims for the Taung child. Of those scientists who had seen and studied the original skull, Robert Broom was the earliest – or one of the earliest – to propose in his 1933 book, *The Coming of Man: Was it Accident or Design?*, that *Australopithecus* should be classified as a hominid. This view did not gain general acceptance until after 1950.

In the interim, the claims made about *Australopithecus* were received with derision, astonishment, hostility and all manner of objections. For 25 years after the announcement of the discovery in 1925, the place of *Australopithecus* was in dispute. By the 1950s, however, Dart had been thoroughly vindicated, and in 1984 a prominent periodical (*Science*) included the Taung child and what Dart made of it as one of twenty scientific discoveries that had shaped the life of human beings in the twentieth century. Perhaps the most significant feature of Dart's breakthrough lay in his willingness to overlook the small brain size of *Australopithecus*. Brain size alone, he declared in a 1956 paper, was not the acid test; the form of the brain was more important. Dart had shown that the principle of mosaic evolution had applied to these early claimants to human ancestry – that is, that some parts of the body had hominised in advance of other parts. He had shown, too, that the particular pattern of mosaicism in *Australopithecus* was totally at variance with that prognosticated by his old mentor, Elliot Smith, and by Arthur Keith (1866–1955), who held that brain enlargement must have occurred in the vanguard of hominid evolution: instead, in *Australopithecus* absolute brain enlargement was hardly evident, whereas dental and postural hominisation were.

Another revolutionary idea flowed from Dart's fertile and imaginative brain in the 1950s. He proposed that before the Stone Age, there had been a Bone Age – an era in which *Australopithecus* had used the bones, teeth and horns of prey animals as implements. Dart's 'Osteodontokeratic Culture' (Bone, Tooth and Horn Culture) was based on this concept. This novel hypothesis arose from his study of thousands of fossilised, broken bones of antelopes and other mammals in the *Australopithecus*-bearing cave of Makapansgat in the Limpopo Province, South Africa (see Chapter 6).

Once again Dart's claims were controversial and a new wave of world interest in and argument over the australopithecines was aroused. I realised from my discussions abroad that the world's top scientists did not like to be browbeaten in the way in which Dart was wont to lambast his students. I once made so bold as to gird up my loins and raise the matter with Dart privately in his office. I referred to his tendency to hammer his fellow scientists, to overdo the text and the illustrations, with far too many specimens in each, and to my belief that his tendency to overstate the case, was putting too many people off. He looked at me, not unkindly but bridling a little, and said, 'Phillip, I have to do it this way, with such a new and revolutionary concept.' He warmed a little: 'If you don't give the [expletive] 200 per cent, they [expletive]-well won't believe the half of it!' The frown gave way to a sudden warm smile and chuckle.

Dart's osteodontokeratic claims found relatively few supporters. Following the work of mainly Charles Kimberlin ('Bob') Brain, many scholars now see the special features of most of the Makapansgat bones as the consequence, in the main, of non-hominid activities (big cats, hyenas and porcupines were probably the principal bone-breakers and accumulators). However, that is less important than the fact that Dart's hypothesis and his insistent advocacy of it spawned a new series of inquiries that were quite fundamental to the study of fossils. When he first proposed his bone-tool theory in the mid-1950s, scarcely anything was known of what other animals, as well as agencies such as the sun and water, did to bones after death. To disprove Dart's proposal, a host of new studies were made on what happens to bones after death.

Once again, those imaginative, fanciful and innovative qualities in Dart's make-up, as well as his willingness to free his mind from the shackles of authority and what is often called 'conventional wisdom' – those very qualities against which Sir Arthur Keith reacted over 60 years earlier – confounded Dart's critics by generating a new discipline, taphonomy.

DART WAS A man rich in idiosyncrasies, a born actor with overwhelming charisma. It could not be claimed that he was a good didactic, systematic teacher and I never learned any factual anatomy from him. But was that the acid test of a great professor? His histrionics were extraordinary and memorable. His unconventional methods in the laboratory earned him the nickname of the Terror of the Dissection Hall, but his methods taught his students the significance of skill (see Chapter 4).

Soon after I had joined Dart's full-time staff in 1951, he took me by the arm and steered me into a crowded dissection hall. 'Stop!' he shouted. It was the story as before and included the hurling of a three-legged stool across the hall. For the life of me I cannot recall what message this action was designed to convey. The usual tirade followed on skill and poise, and the inability of people to control their movements after a million years of evolution! He alternated between pallor and a hectic flush, his voice soaring magnificently, until I feared for his health – then he again took me by the arm and led me out of the hall. As the doors closed behind us, he turned to me, gave a stage-wink and then vented an explosive, hysterical laugh, slapped his thigh and said, 'Got to show them a thing or two now and again!'

Dart's histrionics were also on display in the Senate of the university, which is the supreme governing body on matters academic. Dart not infrequently gave vent to outbursts of anger in which he castigated the Senate, and sometimes its chairperson who was the university's principal and vice-chancellor. Then, seemingly in high dudgeon, he would flounce out of the Senate chamber and not return for the remainder of that meeting. Later, he would slip into the office of Val Greathead, the minutes secretary – and ask her, 'How was I? Was it good?'

The more closely I came to know him, the better I was able to dissect those qualities that made him unforgettable. Dart's stimulating spirit has been the most memorable feature of the impact he has made on those about him. He would never fail to arouse and maintain enthusiasm, whether he was expounding upon the morphology of the nervous system to a class of anatomy undergraduates, or discussing with a junior technical assistant the problems of using a soldering iron.

In my Honours, Masters and Ph.D. years I fell resoundingly under his spell. Was it his research prowess? We knew of the recognition by Dart of the little ape-man skull of Taung, the type specimen of *Australopithecus africanus*. We knew how he had stuck to his guns for decades in the face of nearly universal opposition. Such happenings bring sweat, tears and isolation to the discoverers.

I experienced this for myself in 1964 when, with Louis Leakey of Nairobi and John Napier of London, I was one of the triumvirate of scholars who recognised that some hominid specimens, which Mary and (eldest son) Jonathan Leakey had found at Olduvai Gorge in Tanzania, did not belong to any of the known species of the human family. Instead, we created a new species for which we turned to Dart for a name. Over the weekend I hoped against hope that he would not come up with one of his beloved polysyllabic names. To our delight, he invented the simplest, shortest name he had ever concocted: *habilis*, meaning skilled or dextrous – for I had told Dart that this creature was probably the manufacturer of the most ancient, Oldowan stone tools. Appropriately *Homo habilis* was the earliest 'handyman'.

Again, the top figures scorned our proposal that we published in *Nature* in 1964. Even before we issued this preliminary account of the new Olduvai specimens, the news had spread that I was busy analysing these new specimens. Already I was able to show that their brains (or more precisely their endocranial capacities) were some 45 per cent larger than those of *Australopithecus africanus* and their teeth a little smaller and narrower than those of the South African early hominids. Everybody seemed to know that Louis Leakey was convinced, as by instinct, that we had a new species. I was more cautious and had not yet committed myself. At that time Sir Wilfrid LeGros Clark of Oxford wrote begging me not to allow Leakey to talk me into making a new species to accommodate the Olduvai hominids. Sherry Washburn made similar entreaties to me. I had to reply that I would not allow myself to be talked into anything by Louis Leakey: what would persuade me one way or the other would be the morphological evidence of the small but growing sample of the odd little people of Olduvai. After our triply authored paper appeared, Leakey and Napier moved off in other directions and I was left carrying the habiline banner in the face of criticisms. It was another instance of a premature discovery. While Dart's measure of prematurity had been 25 years, mine was 15 years.

My experience with *Homo habilis* gave me a vivid conception of what Dart had gone through and of the long, lonely years of rejection. Dart's ability to confront and surmount these difficulties reflected his dogged staying power, rather than any heightened scientific insight.

Mind you, it was not Dart who first set my feet on the pathway of research. I owe that influence to Joseph Gillman, in the years when he was a senior lecturer and then reader in the Anatomy Department. Gillman's provocative and forceful style of argument, challenge and leadership in science, his willingness to swim against the current, brought out the best in many people, myself included.

In Dart's honour, I was moved to edit no fewer than three *festschriften* – in *The Leech*, when Dart retired from the chair at 65; in the *South African Journal of Science* to celebrate his 75th birthday; and in the *Journal of Human Evolution* on his 80th birthday. Three honorary publications in tribute to the same man and edited by the same protégé – that is surely unusual. It was a measure of my affection for this father-figure.

In the midst of a committee meeting late one November afternoon in 1988, I received a phone call from a hospital in Sandton that Raymond Dart was sinking. Still badly shaken by the death of my 97-year-old mother in Durban ten weeks earlier, and in a deeply emotional state, I asked one of my staff, Jeff McKee, to take me out to the hospital without delay. Raymond was unconscious though he did respond slightly to my voice close to his ear. It was clear that his end was near and he died in Sandton on 22 November 1988 at 95 years of age.

Raymond Dart's impact was partly the force of his personality. William W. Howells, a distinguished professor of physical anthropology at Harvard, remarked: 'Dart has been called many names in his long joint career with the Taung child ... Phillip Tobias has called him ebullient, emotional, imaginative, charismatic and also, although inclined to scientific heresies, perceptive and prescient. How very true. Let us add, tenacious and courageous.'[13] Above Dart's personality traits, however, there rises the image of a scientific missionary. The revolution he wrought in the knowledge about human evolution, Dart's Taung discovery and what he made of it, will be remembered as the most fundamental single breakthrough in the history of palaeo-anthropology. Next to it the labours of all those who have come after him have simply filled out the details of man's tortuous path of development over the last six or seven million years.

30 | The Darwin of the twentieth century

Another icon of twentieth-century science who played a formative role in my development and thinking over the years – even if at more of a remove than Raymond Dart – was the great Russian-American geneticist Theodosius Dobzhansky ('Doby').

Doby's intensive researches of the early 1930s led to his being invited to deliver the Jessup Lectures at Columbia University in 1936. These lectures resulted in the first edition of *Genetics and the Origin of Species* (1937), perhaps his most important single contribution to science. This modern synthesis of evolutionary theory has been regarded as the twentieth-century counterpart of Charles Darwin's *The Origin of Species* (1859). More than any other, this work earned for Dobzhansky the title of the 'Darwin of the twentieth century'. He synthesised in it all that had been learned since Darwin had written *Origin* and, especially, since the dawn of modern genetics in 1900.

At one and the same time, this book consolidated a new branch of scientific investigation and established the neo-Darwinian synthesis on a secure basis. Dobzhansky's researches led him out of the laboratory into the field. His book of 1937 heralded what L.C. Dunn called, in the Preface, the Back-to-Nature movement. Dobzhansky's work ushered in this movement in two senses: not only did it betoken what Dunn intended, namely the return of geneticists from 'an apparently narrow alley, hedged in by culture bottles of *Drosophila* and other insects' to that 'ultimate laboratory of biology, free nature itself'; but also, in this return to the organic world, the unity of nature engendered anew the holistic approach and the specialist became once again the biologist.

I have already detailed (see Chapter 5) how, when I read my copy of *Genetics and the Origin of Species*, I was smitten by the breadth of Doby's vision and became determined to colour my life's work in comparable shades of genetics and evolution.

Then, Doby was one of the external examiners of my Ph.D. thesis (see Chapter 11), and he was kind enough to write the Foreword for my first book, which was based on this thesis (see Chapter 12).

It was, however, only in 1968, on my second visit to America, that I called on Doby in his New York laboratory, and a lifelong friendship across the hemispheres sprang from that first personal encounter. I found him in the Rockefeller University (previously the Rockefeller Institute). There the superb facilities were arranged to provide men and women of genius with every opportunity for research, free from the responsibilities of formal teaching, unhampered by interference and with a minimum of administration. He was heavily involved in experiments on the genetics of behaviour.

Doby was a shortish, rather stocky man, his white hair stiff and standing on end like the bristles of a bottlebrush. He had a high, somewhat reedy voice, still strongly accented 50 years after he had left the Soviet Union. Yet withal, he spoke and wrote lucid, perfect English. He prided himself on his mastery of several other tongues and, when invited to congresses in foreign countries, insisted on lecturing in the language of the hosts.

When Doby took me home that evening to his apartment I met his wife, Natalia, who had herself been a geneticist when they married in 1924. Her warm personality and hospitality pervaded that pleasant, relaxed, social evening in their New York apartment. I found a friendly atmosphere and stimulating conversation on a wide variety of topics awaiting me. Their home was filled with Russian Orthodox icons: collecting them was obviously one of Doby's lifelong hobbies.

Although Doby expected to continue working at Rockefeller after his formal retirement, a change of policy by the Foundation led to the disbandment of his department. This was a great blow to Doby, just at the time (1969) when he suffered the loss of his beloved Natalia. A sadder man, he moved to Davis, California, where in 1971 a position was found for him as adjunct professor. His former student, Francisco José Ayala, became professor of genetics in the same year. There it was that Doby built a house and spent the remaining four or five years of his life.

IN 1971, DOBY and I were participants in an international colloquium of Accademia Nazionale dei Lincei, Italy's leading and the world's oldest scientific academy, on the subject *L'Origine dell 'Uomo*. The meeting had been organised to commemorate the centenary of Charles Darwin's *The Descent of Man* (1871). As we walked through the streets of Rome together on several evenings, looking for pleasant restaurants at which to dine, we planned his trip to South Africa.

For several years, I had been trying to persuade Doby to accept an invitation from the Institute for the Study of Mankind in Africa, based at Wits in Johannesburg, to come to South Africa to deliver a Raymond Dart Commemorative Lecture. At first he demurred because of South Africa's apartheid policy. With his strong anti-racist stance, he believed he should not come to South Africa until the racist and discriminatory policy of apartheid had been overthrown. I fully understood and sympathised with his standpoint, but I tried to persuade Doby how much more effective and impressive it would be if he who was rated as 'one of the greatest geneticists in the world', were to deliver in Johannesburg a powerful, scientific counterblast to the racial assumptions that underlay South Africa's policies. A man with the universal and anti-racist outlook of Dobzhansky should see for himself the horrors of the repressive political system. He might also have become aware that there were pockets of opposition and resistance to these deplorable policies in the English-medium universities (most notably Wits and Cape Town), certain churches (particularly the Catholics, Anglicans, Methodists, and others) and many, though not all, of the English-medium press, some of which gave a shining lead (as long as they were permitted to) in opposing the policies under which the country writhed for over 40 years.

Eventually, Doby agreed to come as long as he would be at liberty in his Dart Lecture to speak freely against the evils that the twentieth century had perpetrated in the name of race. I was living in a country in which, under the apartheid regime, every white South African considered him- or herself an expert on the subject of race. It might do much good for such a man as Dobzhansky to tell South Africans, including their government, about the correct meaning of race. He had never been to South Africa and his love of travel was as keen as ever.

Alas, more than three years were to elapse before the idea could be effected. With the generous assistance of the Students' Visiting Lecturers Trust Fund, I was able to extend an invitation to him to visit Wits University. He replied to me on 13 February 1975 and his letter was both realistic and prophetic:

> Such a visit would be [of] immense pleasure to me, as well as an opportunity to learn many things heard (and unheard) about by personal experience. Let me say at once that the invitation will be gladly accepted, if I am still alive and in reasonable health. For a 75-year-old (76 years old in 1976) such a proviso is not a result of undue pessimism!

I was not to know then that he was suffering from leukaemia and had but ten months to live. In September 1975, he was laid low by an attack of meningitis. Before the month was up he was out of hospital, actively planning a week's trip to Mexico in December and his visit to Africa, counting fruit-flies of the genus *Drosophila*, and writing chapters for a new book on evolution jointly with Francisco Ayala, Ledyard Stebbins and James Valentine (a book that appeared only in 1977, over a year after his death). Added to all of this, he was beginning to write his Raymond Dart Lecture. In one of his letters to me, telling of these activities, he rhapsodised, 'What more to keep an old man happy!'

Dobzhansky's work had been widely recognised and acclaimed by the award of no fewer than 21 honorary degrees from universities all over the world. This is probably a record for any biologist in recent times, Had he lived to come to Wits, I have little doubt that a 22nd such honour would have been added to this list: on 17 December 1975, Professor Hugh Paterson (head of the Zoology Department at Wits) and I signed a long letter to the university proposing that Doby be awarded an honorary doctorate of science. Little did we realise that that very day was the eve of the end of his life.

His death not only robbed the world of one of the really outstanding evolutionists of the twentieth century; it denied Johannesburg the opportunity of hearing the 13th Raymond Dart Lecture from his lips. Moreover, it precluded his planned sojourn at the University of the Witwatersrand as a visiting professor in the Departments of Anatomy, Genetics and Zoology in February and March 1976. On Wednesday, 17 December 1975, he had been working on his Raymond Dart Lecture in Davis, California. The half-completed manuscript of the address he had proposed to deliver, entitled 'Humankind: A Product of Evolutionary Transcendence', was found on his desk.

As already mentioned, Francisco José Ayala, one of Doby's protégés, headed the Department of Genetics of the University of California at Davis, where Doby had been based since 1971. Ayala was the author of more than 100 scientific and philosophical papers and had worked closely with Dobzhansky, co-authoring books and papers with him. Consequently, the Wits-based Institute for the Study of Mankind in Africa felt it appropriate to invite Ayala to complete Doby's unfinished lecture. This he willingly undertook, using Dobzhansky's notes and passages from previous publications. His aim, Ayala professed, was to 'remain as faithful as possible to Dobzhansky's ideas and even his own words'. The lecture based on that combined authorship was duly delivered at a Dobzhansky Memorial Evening organised by the Institute on 19 July 1978 in Johannesburg. I presented a memorial address on the life and times of Dobzhansky, while Trefor Jenkins (the professor of human genetics at Wits) read the Dobzhansky-Ayala Lecture.

THROUGHOUT HIS CAREER, Dobzhansky had shown an interest in the broader philosophical implications of his work. With the rise of Lysenko in the Soviet Union, and of Nazi ideas on race, he wrote frequent reviews of works in these fields in an attempt to arouse his fellow scientists to the dangers. He was led to write such books as *Evolution, Genetics and Man* (1955) and *The Biological Basis of Human Freedom* (1956) 'to show the bitter, old and wrinkled truth, stripped naked of all vesture that beguiles'. Gradually he approached the genetics and evolution of humans. His Silliman lectures at Yale led to his remarkable synthesis, *Mankind Evolving* (1962), in which he drew together much that was known of the genetics of humans, physical and cultural anthropology. In matters of general ethical moment, there followed *The Biology of Ultimate Concern* (1967) and *Genetic Diversity and Human Equality* (1973).

Of his *Biology of Ultimate Concern*, Doby was a little coy. Indeed, sitting chatting in the Accademia dei Lincei in Rome in 1971, he commented – in reply to my query about this work – 'Of all the books I have written, I am least proud of this one!' This diffidence when the scientist ventures out of his science into the realm of faith is referred to by Doby himself in the Preface to this book, when he says:

> Whatever expertness I may possess is in biology, more precisely in evolutionary genetics. This is no warrant for embarking on speculations in the realms of philosophy and religion. Such speculations are often regarded, among scientists, as regrettable foibles or even as professional misdemeanours. They are as often as not kept secret, for being caught at them is liable to damage a scientist's reputation.

Yet, having stated this unpalatable fact of life, he goes on to do just this: he tries to include biology in a *Weltanschauung*: 'biology is relevant to philosophy. Perhaps it is even more relevant than most other sciences'.

To give some idea of Doby's thinking on these sublime matters, I quote a few lines from the opening of *The Biology of Ultimate Concern*:

> Dostoevsky makes his Ivan Karamazov declare: 'What is strange, what is marvellous, is not that God really exists; the marvel is that such an idea, the idea of the necessity of God, could have entered the head of such a savage and vicious beast as man; so holy it is, so moving, so wise, and such a great honour it does to man.'

Dobzhansky comments:

> This is even more marvelous than Dostoevsky knew. Mankind, *Homo sapiens*, man the wise, arose from ancestors who were not men, and were not wise in the sense man can be. Man has ascended to his present estate from one still more savage, not necessarily more vicious, but quite certainly a dumb and irrational one. It is unfortunate that Darwin has entitled one of his two greatest books the 'Descent', rather than the 'Ascent' of man. The idea of the necessity of God and other thoughts and ideas that do honour to man, were alien to our remote ancestors. They arose and developed, and secured a firm hold on man's creative thought during mankind's long and toilsome ascent from animality to humanity.

> Organisms other than men have the 'wisdom of the body'; man has in addition the wisdom of humanity.

Was it this seldom revealed facet of Theodosius Dobzhansky that the senior American palaeontologist, George Gaylord Simpson, referred to, when, at Davis, he entitled his contribution to the Dobzhansky Memorial Symposium, 'A new heaven and a new earth – and a new man'?

Dr Bob Brain of Pretoria visited Dobzhansky six weeks before his death. Through Bob's kindness I have a lovely photograph of Doby: on the wall behind him are two pictures, iconic of his life's work. One is a portrait of Charles Darwin; the other is a picture of *Drosophila*, the fruit-fly to which so many of Doby's researches were devoted. Together they symbolise, 'Genetics and the Origin of Species'. I feel privileged that my intermittent association with Doby blessed me with a valuable friendship and a major influence on my life.

31 Travelling fossils

It is very exceptional and unusual for original fossil hominid treasures to be put on display. These specimens are not merely rare, but unique. Instead, beautiful casts, accurate as to size, shape, detail and colour, are available to view in museums, university departments and institutes. For a number of years Wits has had a full-time casting officer. At present it is Peter Montja and his predecessor was Eriam Maubane. The products of their handiwork are superb and are on display all over the world.

A specimen such as the Taung child skull, for example, is irreplaceable. Historically, it provided the world's first striking evidence that the early hominids had evolved in Africa and what manner of creatures our early ancestors were. Just imagine the calamity if, to satisfy the needs of an exhibition in another country, harm, damage, loss, and/or destruction, befell the specimen. Of course it could be argued that such damage or disaster might overtake the fossil even in its country of origin. This is true, and it is precisely why we are so careful in prescribing conditions and rules for access to our fossil hominids even on home base.

These were the thoughts running through my head in the early 1980s, when the American Museum of Natural History sent letters to a number of palaeo-anthropologists who were custodians of fossil hominid specimens or collections. To counter the so-called 'Creation Science' movement in the USA, which had been disparaging the evidence on which human evolution was based, the Museum planned to mount a powerful counterblast. This would take the form of a display of some of the world's most outstanding fossil hominid specimens. It was to be an exhibition that differed from other museum exhibits of hominid fossils, in that *original fossils* and not simply casts were to be displayed.

When I received Eric Delson's letter inviting me to bring the Taung skull (and one or two other specimens) to the proposed 'Ancestors Exhibition' in the American Museum of Natural History, New York City, I had first to satisfy my own conscience that it would be worth the risk. Then I had to clear the arrangement with the highest authorities in the university. Only then could I place the matter before the South African National Monuments Council (now the South African Heritage Resources Agency).

While I was pondering the situation, word reached me that there was some resistance from various parts of the world. Coincidentally, most of us who were the interested parties and custodians were invited to attend the First International Congress of Human Palaeontology, at Nice, from 16 to 21 October 1982. At this meeting Delson gathered all those custodians present and appealed to us to support the project in the interests of showing the global community just how voluminous was the available stockpile of the world's fossil hominids. Then trouble broke out.

Michael Day (from London) was disdainful of the plan. How could we expose our precious fossils to the extreme risks involved? Day added, 'Besides, we in England don't have a "Creation Science" movement of any note or consequence. Why should we expose our fossils to such severe risks, for the sake of countering the problem in the USA?' The Chinese, with their wealth of newly found 'Peking Man' and other kinds of fossils were lukewarm. Tanzania, which possessed all of those marvellous fossils of Olduvai Gorge, and those others of Garusi and Laetoli, and Peninj in the north, refused to allow its fossils to travel. Some others were doubtful or negative: they could not see what their own countries would stand to gain, and they were unwilling to take the risk.

It was at this point in a rather heated altercation that a new thought occurred to me – which, for most countries, saved the day. If it was only for the sake of exhibiting the fossils in the hope of impressing the American public and giving a countermove against the American anti-evolution movement, it was clear that the rest of the world would not be won over. Another purpose was clearly required for the American Museum plan to succeed. I myself felt I needed a stronger case before I could with a clear conscience take the proposal to my university and to the National Monuments Council.

The proposal I came up with was based on two previous instances: one featuring Weidenreich and Koenigswald and their fossil hominids from

China and Java respectively; the other involving Koenigswald and myself and our Javanese and East African fossil hominids.

IN 1938, GUSTAV Heinrich Ralph von Koenigswald was working in Java. He had been understandably put out by bitter and intemperate attacks, led by Eugène Dubois, on himself and his interpretation of the fossils he had recovered from Sangiran in eastern Java, especially the second cranium Koenigswald had obtained from Sangiran. The new Javanese cranium had yielded a particularly fine endocranial cast. This was studied by two Netherlands neuro-anatomists, C.U. Ariëns Kappers and K.H. Bouman. They showed that the endocranial cast of the first Javanese calotte, that of Trinil, which Dubois had found in the early 1890s, was concordant with that of Koenigswald's Sangiran cranium of the 1930s. Dubois would not accept that the Sangiran specimens were of the same species as Trinil and his opposition to the young Koenigswald was fierce and bitter.

In the gloom generated by Dubois's onslaught, Koenigswald was cheered and encouraged by a visit – the second he had paid – by Teilhard de Chardin. The latter had been working at Zhoukoudian, the discovery site of 'Peking Man'. Teilhard's visit was propitiously timed to direct Koenigswald's mind to the relationships between Java Man and Peking Man.

A direct comparison of the two sets of Asian remains was essential in order to confirm or refute the relationships. Consequently, Koenigswald readily accepted the invitation of Franz Weidenreich to go to Peking (it was not yet called Beijing) with the new Sangiran cranium. At the beginning of 1939, this historic meeting took place at the Rockefeller All-Union Medical Centre in Peking. Koenigswald describes the scene:

> We laid out our finds on the large table in Weidenreich's modern laboratory: on the one side the Chinese, on the other the Javanese skulls. The former were bright yellow and not nearly so strongly fossilized as our Javanese material; this is no doubt partly owing to the fact that they were much better protected in their cave than the *Pithecanthropus* finds, which had been embedded in sandstone and tufa. Every detail of the originals was compared: in every respect they showed a considerable degree of correspondence ... The two types of fossil man are undoubtedly closely

> allied, and Davidson Black's original conjecture that *Sinanthropus* and *Pithecanthropus* are related forms – against which Dubois threw the whole weight of his authority – was fully confirmed by our detailed comparison. Unfortunately, our carefully documented joint statement could only be published in a much abridged form, on account of the war that had meanwhile broken out in Europe.[14]

In 'lumping' the Chinese and Javanese forms into the prehominids (a category or stage suggested earlier by Boule), Koenigswald and Weidenreich foreshadowed the later decision of palaeo-anthropologists to sink both genera, *Pithecanthropus* and *Sinanthropus*, in the genus *Homo*, as the species *Homo erectus*.

That remarkable Sino-Javanese fossil gathering proved the undoubted value of comparisons being made between original specimens. The two men sitting amid the original specimens in Peking were able to perceive resemblances and parallels between the two regional samples and so to understand their probable relationships to one another and to the general pattern of hominid evolution. This meeting in 1939 must have provided Koenigswald with the inspiration for another such meeting a quarter of a century later between himself and myself and our respective Asian and African fossil samples.

AS I HAVE already mentioned (in Chapter 12), during 1964 I spent a term as a visiting professor at Cambridge University in England. Sadly, Jack Trevor's health was in rather serious decline and he was unable to fulfil his lecturing commitments. Consequently, the Faculty of Archaeology and Anthropology invited me to give a course of lectures in his stead and Wits gave me special leave to accept.

It was an exciting time in the history of our studies of the Olduvai fossil hominids and Louis Leakey, John Napier and I had published our joint article in *Nature* on 4 April 1964, in which we had created the new species, *Homo habilis*.

Louis and Mary Leakey arranged for the original Olduvai fossils and the Peninj mandible, all from Tanzania, to be brought across to England, so that I could continue my detailed study and analysis of them whilst working at

Cambridge. They were deposited by Mary Leakey in the vault of her bank in London. The bank manager was authorised to hand them over to me. Before I agreed to take delivery of them, I wished to check every item against the list I had received from Louis. An amazing spectacle ensued as I spread all of the wrapped specimens over the carpeted floor of the panelled office of the manager. The fossils were wrapped in yards and yards of green toilet paper and I carefully extracted each one of them in order to perform my inventory. 'I say,' said the bank manager, 'I must show this sight to my deputy.' Soon all three of us were on all fours on the carpet, whilst I identified each specimen to the flabbergasted senior officers of the bank. Then, once I was satisfied that nothing was missing or damaged, I retrieved the green toilet paper, wrapped each fossil up, packed them all in the suitcase in which they had travelled to London, and took off for Cambridge. The bank sent me in one of its special vehicles, with an armed guard, to King's Cross Station and saw me safely on to the train. In sharp contrast to my departure from London, my arrival in Cambridge was low-key and unprotected; nobody was there to meet the fossils or me. Only the next morning was I able to unpack the fossils in the Duckworth Laboratory and place them in a special safe that had been put in the laboratory for the purpose.

For the next few months, the table top in Jack Trevor's office was daily bestrewn with the superb East African fossils and people came from far and wide, to see the specimens.

It was at this juncture that I received a letter from Koenigswald inviting me across to Utrecht with all of Leakey's African fossils. He rightly believed that it would be an invaluable opportunity to compare the original specimens of South-East Asia (which he had brought to Utrecht) with those from East Africa. I consulted Louis Leakey by mail and he was emphatically unwilling for the fossils to be carried across to The Netherlands. Undaunted, Koenigswald called me to say that he would then come to Cambridge with all of his Javanese fossils (which were already well travelled).

For days on end we sat together in Jack Trevor's office, with all of the East African and Asian original fossils spread out on the table before us. It was the first time that such an Afro-Asian ingathering of fossil hominids had taken place. I called the meeting 'an Afro-Asian meeting – with a difference'. We spent hours excitedly comparing specimens directly, measuring, describing, annotating, photographing and discussing. No two schoolboys poring over a secret find could have been more fervent than the pair of us.

My dear friend Shirley Coryndon, formerly of Nairobi, and a palaeontologist in her own right, helped us by taking notes as we made observations and took measurements. Eventually, we wrote a joint paper in which we drew attention to the resemblances and parallels between the various grades of hominisation represented in the African and the Asian fossil sequences. Our paper appeared in *Nature* the same year.

Many photographs were taken of the Afro-Asian colloquy. One that I greatly prize is a picture taken on a day when Jack Trevor was well enough to join us in the laboratory: it depicts Koenigswald, Trevor and myself and a table crowded with our very old friends from Java and Tanzania.

THE ARGUMENTS OVER Delson's and the American Museum's proposals continued in 1982 at Nice and some withdrew from the discussions before a decision had been reached. The two precedents for the movement of hominid fossils of immense value, one intra-continental and one intercontinental, were very much in my mind. Both of those precedents had enabled the scientists concerned to compare their respective sets of fossils, with valuable results for the science of human palaeontology. If the proposed ingathering in New York City could provide such opportunities for the scientist-custodians, I reasoned it would make the risky project worthwhile and would be likely to win over some of the most obstinate of the opponents. Consequently, I proposed to the informal meetings in Nice that, one or two weeks before the proposed exhibition opened, the fossils should be made available in a separate and secured study room, under guards. Authorised scientists alone were to be admitted and good lighting and measuring instruments were to be provided. The visiting scientists would be allowed to take photographs of individual fossils. Each evening the fossils were to be returned to the strongroom. The proposal won the day. Some even came to see the scientific study facilities to be the main objective, while the exhibition served a supplementary purpose. With this objective in mind, a number of the erstwhile adversaries withdrew their previous resistance and agreed to support the plan.

I now felt free to embark on the project and was able to persuade Wits and then the National Monuments Council to agree. The workshop in the Wits Anatomy Department made up a strong, but lightweight 'carrycot' for the Taung baby and its relatives. In crossing the Atlantic from the west African coast to John F. Kennedy Airport, we had one of the most turbulent flights I can recall. My anxiety over the 'child's' state of health allowed me very little sleep. At JFK, I was met by security guards (the American Museum had left no stone unturned), taken through the formalities ahead of the other passengers and into a 'stretch' limousine with darkened windows. On the back seat sat the Taung skull's carrycot – and me – together with Dr Michael Cluver, the director of the South African Museum, Cape Town, with one of his fossil treasures, with outriders escorting us. We were rushed through the streets of the awakening city to the American Museum's side-entrance. Again with armed guards covering our arrival, we were taken into a strong-room where I unpacked the carrycot to check that there had been no loss or damage in transit. The specimens were intact and were locked away. Then, and only then, did I succumb to fatigue and nervous strain. I was whisked off to my hotel and 'crashed' (as the Americans say), guarded only by a 'Do not disturb' sign on my door.

The unpacking of the original fossils and their deployment in the study room went off without a hitch. Interested and involved colleagues from many parts of the world busied themselves over the study tables, buttonholing custodians and guards, photographing specimens singly and in pairs or groups, comparing, measuring and taking notes. It was a unique and prodigious occasion and nothing like it had ever been held before. Those who enjoyed the privilege of taking part had every reason to be grateful to the American Museum and especially to Eric Delson who had been largely responsible for bringing it to fruition. Two special visitors to the study sessions were Harry L. Shapiro, who had for a long time presided over the Museum's Anthropology Department, and Raymond Dart. By that stage aged 91, Dart had lost his central vision although his peripheral vision was still rather satisfactory. Many of those studying the fossils were as excited to see Raymond Dart as the fossils. By contrast, Shapiro was a young man at 82.

Then came the tricky business of transferring and mounting all of the specimens for the public exhibition. The arrangement of cabinets had been worked out previously and it remained for the museum staff, collaborating closely with the custodians, to put the specimens safely in position. Special stands had been devised, labels had been drafted and checked and corrected where necessary.

While this was going on, anti-apartheid groups were mounting a demonstration on the opposite side of the road: they were protesting against the inclusion of South African fossils in the exhibit. 'Apartheid fossils not wanted' was the theme of the posters. It was with the most mingled of feelings that I confronted this turn of events. On the one hand, I had every sympathy with the anti-apartheid sentiments of the protestors, and was sorely tempted to cross the road and stand with them – as I had often stood with Wits students and the Black Sash in Johannesburg. On the other hand, the very idea of dubbing the Taung child, who had lived in southern Africa some 2.5 million years before settlers from Europe had even moved in to South Africa, as an 'apartheid fossil', was painful, if not ludicrous. In the event, their protests never gained momentum and they disbanded after a few days.

For six months the exhibition was open to the public and some half a million people paid to see it. The stringent security measures paid off and the Ancestors Exhibition suffered no problems to its end. When the exhibition closed, I sent Alun Hughes, who had carried out most of our excavations at Sterkfontein since 1966, to bring the specimens safely back to their home in the Anatomy Department.

THERE WAS ONE other journeying abroad by two or three of our fossils and that was to Italy. Several Italian colleagues were planning an exhibition of original hominid fossils in the Palazzo Ducale in Venice in 1985. This was to be associated with the creation of holographic images of some of the specimens, like that of the Taung skull which a team from the *National Geographic* magazine, in conjunction with myself, had made at the Council for Scientific and Industrial Research in Pretoria. The holographic image of the Taung child on the cover of *National Geographic* in November 1985 excited wide interest and enthusiasm for the possibilities of the method.

My Italian colleagues advised me not to route my journey through Rome, where there had been some recent security problems, but through Frankfurt. I was met by two burly security guards with bulging holsters who escorted me to the Al Italia first-class lounge. There we spent five or six hours waiting for the Italian direct flight to Venice, thus avoiding Rome and Milan. Our flight was called at just after 2.00 p.m. and I took the carrycot on board with me. At about 2.25 p.m. a bomb exploded in the Al Italia section of the Frankfurt Airport. We knew nothing of it as we were on board our plane, but we heard later that two Australian women had been hurt, one of them fatally, and the building had sustained extensive damage. Taung and I had suffered what airlines call a near miss!

On landing at the Venice Airport, I was escorted with my carrycot to a *vaporetto* or water-taxi that took me to my hotel. A little later, as no one came to fetch me, I took the carrycot and walked through the streets of Venice to the Piazza San Marco and thence into the Doge's Palace. After installing the exhibits and taking part in a media conference, I flew back home via Rome – where I missed another airport bomb by just over 24 hours, but this time it was only myself who was at risk. This exhibition also went off successfully, and once again Alun Hughes, some months later, brought our treasure safely home.

32 | To whom do the fossil hominids belong?

I have described the Wits Anatomy Department as the 'home' of the Taung skull and other choice fossil hominid specimens, but this should not be seen to imply ownership of these specimens. I consider that the skull and other fossils are, at the largest level, world treasure, and at a regional level, South African treasure that has been entrusted to Wits, and specifically to the Department of Anatomy, to serve as the fossils' custodian and curator.

This position has not always been the norm. For example, it seems that Raymond Dart considered that he owned the Taung skull. His personal claim to ownership rested on his extraction of the skull from the breccia received in his laboratory (he had not himself excavated the specimen from the deposits of the Buxton Limeworks), and on his remarkable recognition of the unique and thitherto unprecedented complex of traits that pointed to the child's special place in hominid evolution. Such visitors in 1925 as Robert Broom, Ales Hrdlicka, Alfred Sherwood Romer and the Prince of Wales (later and briefly King Edward VIII) enjoyed free access to the Taung skull, but it remained to all intents and purposes Dart's personal property.

Not only Dart but his university and the Witwatersrand Council of Education were of the impression that he owned the skull. This is confirmed by a passage in his autobiographical *Adventures with the Missing Link*:

> Perhaps, like Davidson Black (who had revealed Peking Man to the world), I should have traveled overseas with my specimens to evoke support for my beliefs, and I was presented with this opportunity. The Witwatersrand Council of Education wrote to say they appreciated that, because of the lack of comparative material in the form of anthropoid skulls of corresponding age, it would be impossible for me to perform a satisfactory monographic study of the Taungs [*sic*] skull in South Africa. The council said they were willing to defray the expenses of my going to

> England for this study *provided I donated the skull to the university*. After careful thought, I decided I could not be bound by such a conditional undertaking, nor was I prepared to absent myself for so long a time from the young department [of anatomy] and my newly established home.[15]

With the Council of Education's offer having been refused by Dart, the Taung skull remained to all intents and purposes his personal property. This position persisted from 1925, the year of the announcement of the discovery, to the end of 1958, when Dart relinquished the chair of anatomy to me. At that stage Dart told me that he was handing the custody of the Taung skull over to me; I was to be the guardian and keeper of the skull, an arrangement that has remained.

Dart's claim of ownership was not unique in those far-off days. When the Florisbad cranium was discovered by Professor Thomas F. Dreyer near Bloemfontein in the Free State province in 1932, the two assistants who helped him in the excavation, A.J.D. Meiring and A.C. ('Hoffie') Hoffman, were not allowed to come anywhere near the Florisbad cranium. According to what Hoffman told me years later, Dreyer, overcome by emotion, hugged the cranium to his bosom, while sitting on a small elevation close to the excavation site, and threw clods at the two young men when they tried to come nearer to see the fossil! Dreyer's possessiveness and 'ownership' of the cranium was evident from that point. Some 22 years later, when the Annual Conference of the South African Association for the Advancement of Science met at the National Museum in Bloemfontein, Dreyer, aged and ill, came from his home to the Museum with the Florisbad cranium to show the participants the important skull.

Another example of claimed ownership was of the Italian fossils of San Felice Circeo (Monte Circeo) and Saccopastore. These splendid Neandertal skulls reposed in the Institute of Anthropology at the University of Rome under its director, Professor Sergio Sergi (1878–1972). Sergi told me that during the German occupation of Italy in the Second World War, he became aware that German officers were seeking fossil treasures for Hitler and that they wished to obtain these skulls. Some of the following story Sergi himself told me during my visit in the 1960s. Further details were kindly filled in by my Italian friends, Giorgio Manzi of Universita di Roma 'La Sapienza', helped by Professor P. Passarello, Professor Aldo Segre and Dr Eugenia Segre-Naldini. In the period between July 1943 and June 1944, a German officer called on Sergio Sergi and asked to see these skulls, probably with the aim of sending them to Germany. Sergi told the officer that the specimens were at that time in Messina, Sicily, where his colleague Landogna was making some special studies on the fossils – or so said Sergi. He knew very well that the American forces had already landed in Sicily, so that even if it was true that the fossil skulls were down there, access to them by Hitler's agents would have been impossible. In fact, Sergi had instructed his technician, Maria Ricca, to take the skulls secretly in an unexceptional shopping basket to a well-known church, the Santa Maria della Pieta in Trastevere, Rome, after a clandestine agreement with the clergy of that parish. The place of safekeeping was below the altar of the church. There the skulls reposed, probably until the allied armies liberated Rome in June 1944. At the end of the war, Saccopastore and San Felice Circeo were safely recovered and restored to the University of Rome.

When Sergi retired from the directorship of the Institute of Anthropology, Roman colleagues told me that he was not enamoured of his successor, Venerando Correnti. Consequently, Sergi removed the Saccopastore and San Felice Circeo skulls to his private apartment in the city. In order to see the skulls, I had to seek an invitation to visit Sergi's apartment. The wizened octogenarian received me warmly when I arrived one morning. He had been studying hairs and proceeded to demonstrate to me several instruments that he had invented. It was a little heavy going. My mastery of Italian was limited and Sergi had almost no English. The hours passed and I had not yet fulfilled the main purpose of my visit, to familiarise myself with the Italian fossil skulls. In the late afternoon, he was reminded that I wished to see the fossils. Into the bedroom we went, where they were kept in hatboxes under the bed with one on top of the wardrobe.

Those fossil skulls were most assuredly Sergi's personal property (he believed); after all, had he not saved them from looting by the German officers – and, for that matter, from the 'clutches' of Correnti?

A fourth example of a palaeo-anthropologist who firmly believed he owned the fossils for which he was responsible was Gustav Heinrich Ralph von Koenigswald, known to his family and friends simply as 'Ralph' (see also

Chapter 31). From 1931–41 he had been responsible for discovering and recovering a number of fossils of 'Java Man' along the Solo River, at the boundary between the middle and eastern thirds of Java in what was then the Dutch East Indies (now Indonesia). These were some of the most important discoveries of *Homo erectus* specimens ever made and contributed appreciably to an understanding of this species' place in time and in hominid systematics and evolution. The details of these admirable fossils need not concern us here.

In December 1941, Japan entered the Second World War. Within days, the famous original fossils of 'Peking Man' had disappeared, while work in Java had come to a standstill. A last-minute American offer to move the original Javanese hominids to the United States was not accepted; in any event, Koenigswald himself apparently did not learn of the offer until after the war. Instead he took extraordinary measures to ensure that the fossils were secreted and protected. Shortly before the Japanese forces occupied Java, plaster casts were substituted for some of the original hominid fossils. In Koenigswald's words:

> The casts were extremely well made and to lay eyes almost indistinguishable from the originals. We had mixed finely ground brick dust with the plaster of Paris, so that even in the event of injury the break would remain nicely dark, as in a genuine fossil. We switched the skulls, so that if the contents of the safe should one day vanish eastwards a few original pieces, at least, would remain in the country.[16]

When the Japanese overran Java, Koenigswald was taken captive and spent many months in a prisoner-of-war camp. However, his beloved wife, Luitgarde von Koenigswald, whom he had married at Bandung in 1935, managed with the help of Javanese friends to stay out of the prison camp. She saved the new Javanese fossil finds, some of which had not yet been described. In this operation, she was helped by two Swiss geologists from the Shell Company, Doctors Mohler and Rothpletz, and a Swedish journalist, Rulf Blomberg. The specimen that Ralph regarded as his most important discovery, namely the upper jaw of Sangiran IV with its large palate and diastema (or space between the upper canine and first premolar), Mrs Von Koenigswald kept in her pocket throughout the Japanese occupation.

Other specimens were concealed by Ralph's friends, the villagers and the 'neutrals'. On one occasion, the Swedish friend, fearing a house search, put the entire collection of isolated teeth that he was safeguarding, including those of *Homo erectus* and *Gigantopithecus*, into large empty milk bottles that he buried in his garden by night.

Because of Koenigswald's foresight, all of the Javanese hominid fossils survived the war. It was a remarkable legacy to posterity and to the post-war flowering of science. His achievement stands in marked contrast to the tragic loss of the Peking Man remains. At the end of hostilities, a weakened Koenigswald was released and he was reunited with his family and all of 'his specimens', save for one of the Solo skulls from Ngandong. This was later found in the Imperial Household Museum of the Japanese Emperor and repatriated to its fellows in Koenigswald's hands.

When the Rockefeller Foundation and the Viking Fund (forerunner of the Wenner-Gren Foundation for Anthropological Research) arranged to bring the Koenigswalds' live and fossil families to America, Ralph had no compunction whatever about packing the Javanese fossils and carrying them to the USA with him.

Then, when the Rijksuniversiteit of Utrecht in the Netherlands created a new chair of stratigraphy and palaeontology especially for Koenigswald, off he went with his itinerant fossils for a twenty-year sojourn in Utrecht. Ralph dreamed of establishing a great international centre for the cherishing and safeguarding of 'his' Javanese hominid fossils and for the study of human evolution, but his plans were not realised in the Netherlands.

Instead, in Germany, the Werner-Reimers Foundation provided the facilities he needed at the Senckenberg Research Institute and Natural History Museum of Frankfurt. Once again, Ralph packed his bags, his fossils and his personal library, and without a 'by your leave', or apparently any consultation with the Netherlands authorities, carried them off to Frankfurt. He still considered them his personal property and, until recently, I knew of only one fossil cranium that he had returned (in 1978) to Indonesia's most eminent palaeo-anthropologist, Teuku Jacob.

TODAY IT IS universally acknowledged that all fossil hominid specimens that are found belong to, and belong in, the country in which they are found. The array of fossils from northern Kenya, which emanated from the east and west of Lake Turkana, and those from the area of the Tugen Hills just to the south, are the property, the national heritage, of Kenya. The Olduvai, Laetoli and Peninj fossils from northern Tanzania are unequivocally Tanzanian treasures. The fossil hominids of Bahr-el-Ghazal and Toros-Menalla in the Chad Republic belong to Chad.

The change came about with the attainment in the 1960s of *uhuru*, independence, decolonisation. Before that, fossils discovered in British, French and German colonies and protectorates were automatically taken to the 'home country'. Kenyan fossils discovered before *uhuru* went to the Natural History Museum in London. The same was true of fossils such as the Kabwe or Broken Hill remains from Zambia (recovered when that territory was still Northern Rhodesia) and the Singa cranium from the Sudan. From Algeria, the hominid fossils of Ternifine, and from Morocco those of Casablanca and Jebel Irgoud, were taken to Paris where they reposed in either the Museé de l'Homme or the Muséum National d'Histoire Naturelle or the Institut de Paléontologie Humaine. There were similar examples from Eyasi in Tanganyika, which were taken to Germany before the Second World War; and from Israel from which fossils were taken to London, Paris and the Peabody Museum of Harvard University.

What should happen to the fossil remains that were removed from the far-flung corners of empires, to various 'homelands' (which of course were not really homelands at all, when looked at from the angle of the fossil human populations)? If it is accepted today by almost all countries and by Unesco that such specimens are part of the legacy of their respective territories of origin, is there any valid reason why this principle should not apply to parts of the heritage discovered when political circumstances were different? To be consistent, the principle should apply retrospectively.

From a purely practical point of view, there are considerable difficulties when a collection of specimens reposes partly in the land of the find and partly in some other centre. For example, someone who wished to study the famous Mount Carmel fossils from the Tabūn and Skhūl caves would have to travel from the Natural History Museum in London, to the Peabody Museum of Harvard University, Massachusetts, to the Rockefeller Museum in Jerusalem. Then, if he or she desired to obtain casts of these fossils, some were officially obtainable from the University Museum of the University of Pennsylvania, Philadelphia.

Next to that wide scattering of the Mount Carmel fossils, the division of the Sterkfontein hominid fossils into over 600 specimens in the School of Anatomical Sciences at the Wits Medical School, Johannesburg, and between 100 and 200 specimens in the Transvaal Museum of the Northern Flagship Institution, in Pretoria, is relatively inconsiderable: the two institutions are only about 50 kilometres apart. That division of the collection founded on historical factors is no serious hardship for the earnest scholar. Moreover, casts can be obtained from the Wits School of Anatomical Sciences, and, latterly, from the Transvaal Museum.

Should there be wholesale repatriation of hominid fossils from their places of enforced exile to their cradle-lands? On grounds of principle, there is no doubt that this would be the most ethical solution, other things being equal. However, we must ask: *are* other things equal? Where we are contemplating the future of objects of such rarity and of such historical and archival world value, we have to ask whether conditions in the source-land are such as to provide adequate protection, security, curatorial skills and custodianship. In many countries such facilities may not be available. This lack would demand help from a body such as Unesco, for the construction of suitable vaults, the provision and training of curators, and the development of a culture of cherishing, appreciating, admiring and valuing the objects in question. Unesco already has such programmes under way in several parts of the world.

AN EXAMPLE OF successful repatriation within East Africa comes to mind. When Mary Leakey discovered in the Olduvai Gorge the magnificent cranium of *Australopithecus boisei* (called originally by her husband Louis Leakey *Zinjanthropus boisei*, and nicknamed variously 'Zinj', 'Dear Boy' and 'Nutcracker Man' based on proposals respectively by Louis, Mary and myself), it was removed to the Leakeys' base at the National Museum in Nairobi. I worked on it there and also in the Wits Anatomy Department for some years. When I had completed my major study, but was still finalising my manuscript, reading proofs, and so on, Louis arranged for 'Zinj' to be returned to the newly independent and renamed Tanzania.

From the scientific academies of Beijing, Moscow, Paris, London and Washington, leading figures in palaeontology came to Dar-es-Salaam for the handing-over ceremony in January 1965. I was invited to be present – but it was stipulated that I attend not as a representative of any organisation or country, but as myself – who happened to be the one who had done the hard work over some five years. Here's the point: a special depository was constructed in the National Museum (formerly the George V Museum) in Dar-es-Salaam. This was made exceptionally secure and was fireproof, with temperature- and humidity-control. It was a model of how the world's most precious fossils should be housed. President Julius Nyerere, known to his nation as 'Mwalimu' or teacher, took a personal interest in the preparation of the repository and played an active part in the ceremony that took place in the grounds of the Museum.

In Nairobi, Richard Leakey (Louis and Mary's second son) was instrumental in having a lavish facility built with generous support from the Royal Swedish Academy of Sciences and other international sources. In this were housed all of the fossil hominids from Kenya that had been recovered after *uhuru*.

Here were two African countries, Kenya a former colony and Tanzania a former protectorate, which had risen admirably to the need for the bones of its earliest citizens to be housed in ideal conditions.

The finest fossil hominid specimen to emerge, until today, from Zambia was the outstandingly complete cranium of what used to be called 'Rhodesian Man' and has since independence come to be known as Broken Hill Man or Kabwe Man. This specimen was recovered by a miner T. Zwigelaar in 1921 long before independence. From colonial Northern Rhodesia, the cranium went to the Natural History Museum in London. There, when I last looked, it still resides. Word reached me that the Zambian government had asked the Museum to repatriate the skull to Zambia. The Museum authorities, for a reason or reasons not entirely clear, apparently declined this request. Of course, Zambia, as such, did not exist when the skull was recovered in 1921. It seems unlikely that the return of the cranium was declined on such legalistic grounds. Was there a fear that facilities in Zambia for storage and curating of the skull might not be adequate; or that, if this specimen were repatriated, it might prove to be the thin end of the wedge leading to a flood of other requests? This is one of the cases known to me where an official, formal request for repatriation of a fossil hominid specimen has been made to a former colonial power by the fossil's source land (and seemingly denied).

There have also been various repatriations to Asia that are worth mentioning. I have already spoken about the return by Ralph von Koenigswald of the calvaria of the Mojokerto child, to Teuku Jacob of Gadjah Mada University at Yogyakarta in Indonesia.

In a recent letter to me, Jacob has thrown more light on repatriations to Indonesia. When Jacob was hospitalised for a few days in Utrecht in 1967, Koenigswald promised him that he would return the collection to Java. In an interview in the *Frankfurter Algemeine Zeitung* in 1974, Koenigswald indicated that he would return the collection to Indonesia. A year later Teuku Jacob brought back the Ngandong skulls. In 1997, he had picked up the Sambungmachan skull 3, which, he wrote, had been 'smuggled away to New York'. Another *Homo erectus* skull had been spirited away and offered for sale by an antique dealer in Switzerland. Sambungmachan 4, he wrote, was now back in Java – in Bandung. All told, according to Teuku Jacob, around two-thirds of the Indonesian *Homo erectus* specimens are now in Yogyakarta. Other pre-war Indonesian hominid fossils are still in Leiden and Frankfurt.

Another small repatriation is worth mentioning. In the 1920s, several teeth of 'Peking Man' from Zhoukoudian Locus A, had found their way to the University of Uppsala in Sweden. In 1978, I was pleasantly surprised to find that the teeth had survived and were still present in Uppsala. I was visiting Uppsala for an historical meeting: it took the form of a Nobel Symposium organised by the Royal Swedish Academy of Sciences to commemorate the 200th anniversary of the death of Carolus Linnaeus (Carl von Linné).

Linné was the great classifier extraordinaire of living things. Fearlessly he braved the wrath of his fellow mortals and claimed that whatever rules were found to apply to other animals should be assumed to apply as well to man. In a sense, Linné brought humans down from the angels to join the apes.

The King and Queen of Sweden joined us for one of the sessions of the symposium and for lunch. I was placed opposite His Majesty at lunch and, catching my eye, he taught me the correct way to drink a *skol*. It was on this occasion that King Carl Gustav conferred on Mary Leakey the Linnean Gold Medal. She was the first female recipient. During a visit to Uppsala, we were shown the Chinese teeth that had been excavated over half a century earlier. During an informal get-together of the symposiasts, Richard Leakey unexpectedly voiced an appeal, on behalf of the international scientists, that the Swedes return the Zhoukoudian teeth to the Institute of Vertebrate Palaeontology and Palaeo-anthropology (IVPP) in Beijing, China. I was in full agreement with the sentiment but felt that the request had been sprung rather gracelessly on our Swedish hosts. However, Richard achieved his aim and the teeth were duly returned. This was not a case of specimens having been taken away before independence, but the removal of the fossils went back to a period when foreigners had had very few scruples about what they did with specimens from far-off cradle-lands.

THE SAD TALE of 'Egbert', the Neandertal youth from Ksâr 'Akil, in the Lebanon, is a story of loss during repatriation. Father Franklin Ewing S.J. had excavated in this cave deposit near Beirut in 1938. With his colleague J.G. Doherty, he recovered human remains said to have been 'neandertaloid' in character. The best preserved was a partial skeleton of a child of about eight years old. Ewing nicknamed it 'Egbert' because (he told me) of the state of preservation of the cranium – 'like a broken eggshell!' When I visited Father Ewing in 1956 (see Chapter 22), Egbert was with him in Fordham Catholic University in The Bronx, New York City. He allowed me to examine and handle it, and I was able to obtain casts of the cranium and jaw. I took photographs of Father Ewing holding Egbert on the steps at the entrance to Fordham University. After Ewing's death, the intention was to repatriate Egbert to the museum in Beirut. The bones were apparently sent from Fordham to the Society of Jesus' headquarters in Austria, with a view to their being returned to Lebanon. The fossil bones have never been seen again, despite rigorous enquiries made by Dr Nancy Minugh-Purvis of Philadelphia and me (as a visiting professor at the University of Pennsylvania in the 1990s). My 1956 photographs of Father Ewing and Egbert may be the last pictures taken of the skull.

The case of two problematical European repatriations is that of the Le Moustier and Combe Capelle remains. The skeleton of a Neandertal youth was brought to light in August 1908, in the Le Moustier cave in the Dordogne District in the south of France. The tools from this cave gave the name Mousterian to the associated archaeological industry. The remains were discovered by a Swiss antiquary, Otto Hauser, who sold the bones to the Berlin Museum. A year later, Hauser recovered another important skeleton, ornamented with seashells, in a bed at Combe Capelle in the Dordogne, France. This, too, was acquired by the Berlin Museum.

These two prehistorical skeletons reposed in the State Museum in Berlin for some 35 years. Near the end of the Second World War, the bombing of Berlin on 3 February 1945 resulted in a direct hit on the Museum. In the ensuing fire, the Le Moustier post-cranial remains, among others, were severely damaged and partly destroyed. As a young student, I grew up on the teaching that the Le Moustier skull had been lost, a casualty of war. In 1957 Boule and Vallois wrote – not entirely accurately – that the Le Moustier skeleton 'was completely destroyed during the last war'.[17] In fact, the Le Moustier skull had been taken to the Soviet Union in 1945! Subsequently the 'lost' and 'destroyed' skull was located in Moscow, whence it was returned – together with the necklace of Combe Capelle and art objects – to the German Democratic Republic in 1958.

Theoretically, if the Soviet scientists had recognised the identity of the skull, they would have been confronted with an unusual dilemma: to which destination should the remains be returned? Should it be to France, from where they were removed almost a century earlier, or to Berlin where, by legitimate purchase, they had reposed since about 1909? By today's thinking, the Le Moustier skull should have been sent to France, but the Russians sent the remains back to Berlin, from where they had presumably been plundered as war spoils.

This must have been an exceptional, if not unique, repatriation quandary, where there were two potential claimants for the 'return' of a fossil *Homo* expatriate.

However, it does not seem that the skulls of Le Moustier and Combe Capelle were ever recognised or closely examined in Moscow. They were returned, along with stolen artworks, in a packing case to Berlin.

In 1997, I was invited to Berlin by the German palaeo-anthropologist, Herbert Ullrich. He gave me the opportunity to examine the somewhat fire-scorched skull of Le Moustier and, by comparing it with earlier descriptions and illustrations, I was able to confirm its identity. Until then, it had been one of very few European hominid fossils that I had not personally examined over the previous 45 years.

At the time of my visit to Berlin in 1997, the skull of Combe Capelle was still 'missing'. Strenuous efforts were made by some German colleagues to find and identify fragments with matching parts in illustrations and measurements that had been published much earlier. Only on 27 December 2001 could Almut Hoffman and Dietrich Wegner announce that they had 'rediscovered' and identified the skull of Combe Capelle without any doubt.

The cases of Le Moustier and Combe Capelle, like those of San Felice Circeo and Saccopastore in Italy, and Ngandong in Java, illustrate dramatically how, under wartime conditions, the purloining – or attempted pillaging – of fossils by invaders may add another dimension to the problems of expatriation and repatriation of fossil hominids.

I staunchly support the idea that fossil hominids should remain in their country of origin, or be restored to it in cases where they have been removed. Where there is doubt about whether the facilities in the cradle-land are adequate, the country of 'adoption', perhaps helped by Unesco, should offer to improve or provide appropriate facilities for the permanent housing of the fossils to be returned. Since this category of heritage treasures comprises very rare specimens, many of which can justly be described as unique, perhaps Unesco should set up a commission to oversee problems of repatriation of fossil hominids, just as there is a special authority to look after sites and collections that have been placed on the World Heritage List.

Against this background, I have always tried to safeguard the precious fossils recovered by my team and me and those entrusted to my care and custodianship. The public is often unaware of just how valuable the original and irreplaceable fossils are. I hope to fulfil this duty until the end of my days.

TALKING OF THAT, as the months pass, my 80th birthday looms ever closer. Life seems to be as busy as ever. At least two or three more books await completion. There are still invitations coming in each year, most of which I cannot dream of accepting, whilst others have so special, so historical a hold on me that I cannot for a moment decline. Several of my friends recently asked me if it wasn't time to slow down, to which I replied that with local and overseas commitments before me, my autobiography and other writing as well as speaking engagements, I just cannot find the time to slow down.

I am not complaining: keeping 'at it' is the way to keep going and to keep young – at least in mind and spirit. I have often told my students: 'Never let your patients put their feet up when they retire. That is surely the kiss of death!' Retirement is for those only who still have things to do – books to read and to write, travels to travel, gentle sports to indulge in and not-so-gentle games to follow in the media, concerts, plays and book launches to attend, time to be spent in nature, grandchildren to look after and entertain, societies and places of contemplation to serve, cricket matches and World Cup games to follow, hobbies to pursue. All these activities help to stave off one's way of life falling into the sear, the yellow leaf (with apologies to William Shakespeare and *Macbeth*), and to ward off 'second childishness, and mere oblivion, sans teeth, sans eyes, sans taste, sans everything' (with thanks to Shakespeare and *As You Like It*).

Much still lies ahead of me: Sterkfontein's ultimate messages, brain evolution, retrospecting in search of synthesis between the persons and cultures of yesteryear and those of today – and prospecting by probing to the humankind of tomorrow, the spread of ancient hominids throughout Africa and thence to the rest of the Old World and the peopling of the Americas and other islands. There is the unfinished quest for synthesis at many levels – genetics and evolution, science and religion, long-term and short-term development, brain, mind and behaviour. To give voice is one matter; to find answers is another. These things and many others must wait for my next volume of memoirs. As you can see, I'm far too busy to slow down now.

Notes

1. Dart and Craig (1959: 83, 90–92).
2. Dart and Craig (1959: 50).
3. Findlay (1972: 65).
4. Findlay (1972: 66).
5. Findlay (1972: 66).
6. Broom (1950: 66).
7. Leakey (1974: 154).
8. Dart and Craig (1959: 58).
9. Coon (1963: vii; emphasis added).
10. Geffen (1991: ix–xiii).
11. Dobzhansky (1973).
12. Steiner (1998: 47).
13. Howells (1985: 24).
14. Von Koenigswald (1956: 101).
15. Dart and Craig (1959: 51; emphasis added).
16. Von Koenigswald (1956: 115).
17. Boule and Vallois (1957: 205 note 22).

Select bibliography

A selection of publications by P.V. Tobias (1945–1965)

1945: Students' scientific expedition to the Makapan Valley. *Wits University Views* [*WU's Views*], Science Supplement: 2
1947: The characterization of the spermatogonial chromosomes of the albino rat (*Rattus norvegicus albinus). The South African Journal of Science*, 43: 312–319.
1947: (with Blom, F., Cohen, S., De Vos, W. and Lunking, F.). Congenital diaphragmatic hernia with special reference to a pericardio-peritoneal hiatus. *The Leech*, 18: 19–22.
1948: Racial discrimination in the medical schools. *Trek*, 12: 18–19.
1948: Danger of state-enforced 'Apartheid' in Union's universities. *The Star*, Johannesburg, 15 December 1948.
1948: Open letter on the threat of apartheid to University Councils, Senates and Academic Staffs, Members of Parliament, other public figures and bodies, from the President of NUSAS [P.V. Tobias]. Cape Town: National Union of South African Students: 1–3.
1949: The excavation of Mwulu's cave, Potgietersrust District. *The South African Archaeological Bulletin*, 4: 2–13.
1949: Medical education in South Africa (joint author). Memorandum of evidence, on behalf of the Students' Medical Council of the University of the Witwatersrand, to the Commission of Enquiry into the training of medical students.
1951: *The African in the Universities. NUSAS Handbook Series*: 4–30.
1951: ed. *Transkei Survey*. Cape Town: NUSAS Research Journal.
1952: The chromosomal complement of the gerbil, *Tatera brantsii draco. The South African Journal of Science*, 48: 366–373.
1953: The problem of race identification: limiting factors in the investigation of the South African races. *The Journal of Forensic Medicine*, 1(2): 113–123.
1953: Trends in the evolution of mammalian chromosomes. *The South African Journal of Science*, 50: 134–140.
1954: Blue sclerae, fragilitas ossium, arachnodactyly and deafness in an Indian family. *The American Journal of Human Genetics*, 6: 270–278.
1954: Climatic fluctuations in the Middle Stone Age of South Africa as revealed in Mwulu's Cave. *Transactions of the Royal Society of South Africa*, 34: 325–334.
1954: Extra-chromosomal chromatin in the germ-cells. *The South African Journal of Medical Sciences*, 19: 57–58.
1954: On a Bushman-European hybrid family. *Man*, 54: 179–182.
1954: The sex chromosomes of the gerbil, *Tatera brantsii draco. The South African Journal of Science*, 50: 228–254.
1955: Physical anthropology and somatic origins of the Hottentots. *African Studies*, 14: 1–22.

1955: The Taaibosch Koranas of Ottosdal: a contribution to the study of the Old Yellow South Africans. *The South African Journal of Science*, 51: 263–269.

1955: Taurodontism in a Bushman skull from Kimberley. *The British Dental Journal*, 98: 352–355.

1955: Teeth, jaws and genes. *The Journal of the Dental Association of South Africa*, 10: 88–104.

1955–56: Les Bochimans Auen et Naron de Ghanzi. Contributions à l'étude des Anciens Jaunes Sud-Africains. *L'Anthropologie*, 59: 235–252, 429–461, 1955; 60: 22–52, 268–289, 1956.

1956: Bushmen of the Kalahari. *Man*, 57: 33–40.

1956: *Chromosomes, Sex-cells and Evolution in a Mammal*. London: Percy Lund, Humphries & Co. Ltd.

1956: On the survival of the Bushmen, with an estimate of the problem facing physical anthropologists. *Africa*, 26: 174–186.

1956: Premeiotic mitosis: a new type of cell-division in mammals. *The Journal of Anatomy*, 90: 570–571.

1956: The progress of desegregation in the United States of America and its lessons for South Africa. *Newsletter* 1 of Education League of South Africa, Johannesburg: appendix.

1958: Growth changes in the occipital bone of Bushmen and Negroids. *The Journal of Anatomy*, 92: 652–653.

1958: Kariba re-settlement: an experiment in human ecology. *The South African Journal of Science*, 54: 148–150.

1958: Proposed establishment of an Institute for the Study of Man in Africa. *The South African Museums Association Bulletin*, 6: 432–434.

1958: Proposed Raymond Dart Institute for the Study of Man in Africa. *African Studies*, 17: 120–122.

1958: Raymond Arthur Dart: biographical sketch and appreciation. *The Leech*, 28: 385–93.

1958: (with Plotkin, R., eds.). *Raymond A. Dart Festschrift*: Commemorative Number on his retirement from the Chair of Anatomy at the University of the Witwatersrand, Johannesburg. *The Leech*, 28(3, 4, 5): 82–155.

1958: Sexing texts and sex-chromosomes. *The Leech*, 27: 7–16.

1958: Skeletal remains from Inyanga. In: R. Summers, ed. *Inyanga, Prehistoric Settlements in Southern Rhodesia*, pp. 159–172. Cambridge: Cambridge University Press.

1958: Some aspects of the biology of the Bantu-speaking African. *The Leech*, 28: 3–12.

1958: Studies on the occipital bone in Africa. Part III: Sex difference and age changes in occipital curvature and their bearing on the morphogenesis of differences between Bushmen and Negroes. *The South African Journal of Medical Sciences*, 23: 135–146.

1958: Tonga resettlement and the Kariba Dam. *Man*, 58: 77–78.

1959: The Nuffield-Witwatersrand University expeditions to Kalahari Bushmen, 1958–1959. *Nature*, London, 183: 1 011–1 013.

1959: Provisional report on Nuffield-Witwatersrand University research expedition to Kalahari Bushmen, August–September 1958. *The South African Journal of Science*, 55: 13–18.

1959: Some developments in South African physical anthropology, 1938–1958. In: A. Galloway, *The Skeletal Remains of Bambandyanalo*, edited by P.V. Tobias, pp. 129–154. Johannesburg: Witwatersrand University Press.

1959: Studies on the occipital bone in Africa. Part I: Pearson's occipital index and the chord-arc index in modern African crania. *The Journal of the Royal Anthropological Institute of Great Britain and Ireland*, 89(2): 233–252.

1959: Studies on the occipital bone in Africa. Part II: Resemblances and differences of occipital patterns among modern Africans. *Zeitschrift für Morphologie und Anthropologie*, 50(1): 9–19.

1959: Studies on the occipital bone in Africa: Part IV: Components and correlations of occipital curvature in relation to cranial growth. *Human Biology*, 31: 138–161.

1959: Studies on the occipital bone in Africa: Part V: The occipital curvature in fossil man and the light it throws on the morphogenesis of the Bushmen. *The American Journal of Physical Anthropology*, 171: 1–12.

1960: Aged Bushmen without high blood pressure. *South African Railways and Harbours Magazine*, October 1960: 115.

1960: *Embryos, Fossils, Genes and Anatomy*. Johannesburg: Witwatersrand University Press.

1960: The Kanam jaw. *Nature*, London, 185: 946–947.

1960: Studies on the occipital bone in Africa. Part VI: The relative usefulness of Pearson's Occipital Index and the occipital chord-arc index. *Man*, 60: 23–25.

1961: An archaeological collection from Vume Dugame on the Lower Volta. In: O. Davies, ed. *Archaeology in Ghana*, pp. 35–38. Edinburgh: Thos. Nelson.

1961: Fingerprints and palmar prints of Kalahari Bushmen. *The South African Journal of Science*, 57: 333–345.

1961: *The Meaning of Race*. Johannesburg: South African Institute of Race Relations, pp. 1–25. (Reprinted in Penguin Books' *Race and Social Difference*, 1972; 2nd ed., revised and enlarged, issued by The South African Institute of Race Relations, 1972).

1961: New evidence and new views on the evolution of man in Africa. *The South African Journal of Science*, 57: 25–38.

1961: Physique of a desert folk. *Natural History*, 70: 16–25.

1961: Studies on Bushman-European hybrids. *Proceedings of the Second International Conference of Human Genetics*, in Rome, 1961, pp. 461–471. Amsterdam: Excerpta Medica Foundation.

1961: The work of the [Witwatersrand University] gorilla research unit in Uganda. *The South African Journal of Science*, 57: 297–298.

1962: Early members of the genus *Homo* in Africa. In: G. Kurth, ed. *Evolution und Hominisation*, pp. 191–203. Stuttgart: Gustav Fischer Verlag (and 2nd ed., 1968).

1962: Glossary of human genetical terms with explanatory notes. *The Leech*, 32: 186–191.

1962: On the increasing stature of the Bushmen. *Anthropos*, 57: 801–810.

1962: A re-examination of the Kanam mandible. *Actes du IVe Congrès Panafricain de Préhistoire et de l'Etude du Quaternaire*. Sections I and II. *Annales Ser. Qu-80, Sciences Humaines*, No. 40, pp. 341–350. Tervuren, Belgique.

1962: Some perspectives in human genetics. *The Leech*, 32: 76–83.

1963: Cranial capacity of Zinjanthropus and other Australopithecines. *Nature*, London, 197: 743–746.

1963: Glossary of terms used in human genetics with explanatory notes. In: *Contributions of genetics to epidemiologic studies of chronic diseases*, pp. 1–18. Published by University of Michigan Medical Center. (Reprinted 1964, 1965, 1966).

1963: Re: Blue-eyed Africans. *The South African Journal of Science*, 59: 127.

1963–64: *Man's Anatomy*, Vol. I (1963), Vol. II (1963), Vol. III (1964), (with M. Arnold). Johannesburg: Witwatersrand University Press. (2nd ed., 1974; 3rd ed., 1977; revised reprint of 3rd ed., 1981, 4th ed., 1988).

1964: Bushman hunter-gatherers: a study in human ecology. In: D.H.S. Davis, ed. *Ecological Studies in Southern Africa*, pp. 67–86, and in *Monographiae Biologicae 14*. (Reprinted in 1968, 1974 and 1976).

1964: The case of *Homo habilis* (with J.R. Napier: reply to B.G. Campbell). *The Times*, London, 5 June 1964.

1964: A comparison between the Olduvai hominines and those of Java and some implications for hominid phylogeny (with G.H.R. von Koenigswald). *Nature*, London, 204(4958): 515–518. (Reprinted 1965 in *Current Anthropology*, and in *L' Anthropologie*).

1964: Dart and Taung forty years later. *The South African Journal of Science*, 60: 325–329.

1964: The evolution of man – reply to B. Campbell and W.E. LeGros Clark. *Discovery*, 25(8): 49–50.

1964: Genetics in medical education. *The South African Medical Journal*, 38: 336–339.

1964: The hominid remains from Olduvai Gorge, Tanganyika. *Proceedings of the Seventh International Congress of Anthropological and Ethnological Sciences*, III: 333–341. Moscow: 'Science' Publishing House.

1964: A new species of the genus *Homo* from Olduvai Gorge (by L.S.B. Leakey, P.V. Tobias and J.R. Napier). *Nature*, London, 202 (4927): 7–9. (Reprinted in 1965 and 1971).

1964: The Olduvai Bed I hominine with special reference to its cranial capacity. *Nature*, London, 202(4927): 3–4. (Reprinted in 1965 and 1971).

1965: *Australopithecus, Homo habilis*, tool-using and tool-making. *The South African Archaeological Bulletin*, 20: 167–192.

1965: (with Blacking, J., eds.). *Bushmen and Other Non-Bantu Peoples of Angola*, by A. de Almeida. Johannesburg: Witwatersrand University Press for the Institute for the Study of Mankind in Africa.

1965: Cranial capacity of the hominine from Olduvai Bed I. *Nature*, London, 208(5006): 205–206.

1965: The early *Australopithecus* and *Homo* from Tanzania. *Anthropologie*, 3(3): 43–48.

1965: Early man in East Africa. *Science*, 149(3679): 22–33. (Reprinted 1966, 1969, 1973).

1965: *Homo habilis*. In: *Britannica Book of the Year 1965*, special report No. 3, pp. 252–255.

1965: *Homo habilis:* last missing link in human phylogeny? In: S. Genoves, ed. *Homenaje a Juan Comas, en su 65 anniversario*, Vol. II, pp. 377–390. Mexico City: Editorial Libros de Mexico.

1965: *Homo habilis* (reply to Ashley Montagu). *Science*, 149(3687): 918.

1965: New discoveries in Tanganyika, their bearing on hominid evolution. *Current Anthropology*, 6(4): 391–399; and reply to comments, 406–411.
1965: The science writer and science. *The South African Journal of Science*, 61: 1–4.
1965: *Zinjanthropus* returns to Tanzania. *The South African Archaeological Bulletin*, 20(77): 1–2.

General publications including a selection of post-1965 publications by P.V. Tobias

Academic Freedom Committees of the Universities of Cape Town and the Witwatersrand. (1974). *The Open Universities in South Africa and Academic Freedom, 1957–1974*. Cape Town: Juta and Co.
Ardrey, R. (1961). *African Genesis*. London: Collins.
Balsan, F. (1952). *L'expédition Panhard-Capricorne*. Paris: Dumont.
Bonin, G. von (1963). *The Evolution of the Human Brain*. Chicago: University of Chicago Press.
Boule, M. and Vallois, H.V. (1957). *Fossil Men*. New York: The Dryden Press.
Brain, C.K. (1958). *The Transvaal Ape-Man-Bearing Cave Deposits*. Pretoria: *Transvaal Museum Memoir No. 11*.
Brain, C.K. (1981). *The Hunters or the Hunted? An Introduction to African Cave Taphonomy*. Chicago and London: University of Chicago Press.
Brenner, S. (1946). The chromosome complement of *Elephantulus*. *South African Journal of Medical Sciences*, Biological Supplement II: 71–78.
Broom, R. (1950). *Finding the Missing Link*. London: Watts.
Broom, R., Robinson, J.T. and Schepers, G.W.H. (1950). *Sterkfontein Ape-man,* Plesianthropus. Pretoria: *Transvaal Museum Memoir No. 4*.
Broom, R. and Schepers, G.W.H. (1946). *The South African Fossil Ape-men: The Australopithecinae*. Pretoria: *Transvaal Museum Memoir No. 2*.
Carter, J.E.L. and Honeyman Heath, B. (1990). *Somatotyping: Development and Applications*. Cambridge and New York: Cambridge University Press.
Centlivres, A. van de S., Feetham, R., et al. (1957). *The Open Universities in South Africa*. Johannesburg: Witwatersrand University Press.
Clark, W.E. LeGros (1955). *The Fossil Evidence for Human Evolution*. Chicago: University of Chicago Press.
Cobb, W.M. (1947). *Medical Care and the Plight of the Negro*. New York: National Association for the Advancement of Colored People.
Cole, S. (1975). *Leakey's Luck: The Life of Louis Seymour Bazett Leakey 1903–1972*. London: Collins.
Cooke, H.B.S., Malan, B.D. and Wells, L.H. (1945). Fossil man in the Lebombo Mountains, South Africa: the 'Border Cave', Ingwavuma District, Zululand. *Man*, 45(3): 10–13.
Coon, C.S. (1963). *The Origin of Races*. London: Jonathan Cape.
Damon, A. (1968). Secular trend in height and weight within old American families at

Harvard 1870–1965, within twelve four-generation families. *American Journal of Physical Anthropology*, 29: 45–50.
Dart, R.A. (1925). *Australopithecus africanus*: the man-ape of South Africa. *Nature*, London, 115: 195–199.
Dart, R.A. (1957). *The Osteodontokeratic Culture of* Australopithecus prometheus. Pretoria: *Transvaal Museum Memoir No. 10.*
Dart, R.A. and Craig, D. (1959). *Adventures with the Missing Link*. New York: Harper and Brothers.
De Villiers, H. (1968). *The Skull of the South African Negro: A Biometrical and Morphological Study*. Johannesburg: Witwatersrand University Press.
Dobzhansky, T. (1937). *Genetics and the Origin of Species*. New York: Columbia University Press.
Dobzhansky, T. (1967). *The Biology of Ultimate Concern*. London: Rapp and Whiting.
Dobzhansky, T. (1973). *Genetic Diversity and Human Equality*. New York: Basic Books, Inc.
Dobzhansky, T. and Ayala, F. (1977). *Humankind: A Product of Evolutionary Transcendence*. Johannesburg: Witwatersrand University Press for the Institute for the Study of Man in Africa.
Eiseley, L.D. (1959). *Darwin's Century*. London: Victor Gollancz.
Findlay, G. (1972). *Dr. Robert Broom, Palaeontologist and Physician, 1866–1951*. Cape Town: A.A. Balkema.
Geffen, L. (1991). Tribute to Phillip V. Tobias. In: P.V. Tobias, *Images of Humanity: Selected Writings of Phillip V. Tobias*, pp. ix–xiii. Johannesburg: Ashanti Publishing.
Gillman, J. and Gillman, T., eds. (1951). *Perspectives in Human Malnutrition: A Contribution to the Biology of Disease from a Clinical and Pathological Study of Chronic Malnutrition and Pellagra in the African*. New York: Grune and Stratton.
Gould, S.J. (1981). *The Mismeasure of Man*. New York and London: W.W. Norton and Company.
Greulich, W.W. (1957). A comparison of the physical growth and development of American-born and native Japanese children. *American Journal of Physical Anthropology*, 15: 489–516.
Greulich, W.W., Crismon, C.S. and Turner, M.L. (1953). The physical growth and development of children who survived the atomic bombing of Hiroshima and Nagasaki. *Journal of Pediatrics*, 43: 121–145.
Hammond-Tooke, W.D., ed. (1971). *The Bantu-Speaking Peoples of Southern Africa*, 2nd ed. London: Routledge and Kegan Paul.
Harrison, G.A., Weiner, J.S., Tanner, J.M. and Barnicot, N.A. (1964). *Human Biology: An Introduction to Human Evolution, Variation and Growth*. Oxford: Clarendon Press.
Hiernaux, J. (1964). Weight/height relationship during growth in Africans and Europeans. *Human Biology*, 35: 273–293.
Hiernaux, J. (1968). Variabilité de dimorphisme sexuel de la stature en Afrique Sub-saharienne et en Europe. In: R. Peter, F. Schwarzfischer, G. Glowatzki and G. Ziegelmayer, eds. *Anthropologie und Humangenetik*, pp. 42–50. Stuttgart: Gustav Fischer.
Holloway, R.L. (1966). Cranial capacity and neuron number: a critique and proposal. *American Journal of Physical Anthropology*, 25: 305–314.

Holloway, R.L. (1975). Early hominid endocasts: volumes, morphology and significance for hominid evolution. In: R.H. Tuttle, ed. *Primate Functional Morphology and Evolution*, pp. 393–415. The Hague: Mouton.

Howells, W.W. (1957). *Variation of External Body Form in the Individual.* Cambridge, Mass.: Peabody Museum.

Howells, W.W. (1985). Taung: a mirror for American anthropology. In: P.V. Tobias, ed. *Hominid Evolution: Past, Present and Future.* New York: Alan R. Liss.

Hughes, A.R. and Tobias, P.V. (1977). A fossil skull probably of the genus *Homo* from Sterkfontein, Transvaal. *Nature*, London, 265: 310–312.

Huxley, J.S. (1963). *Evolution: The Modern Synthesis*, 2nd ed. London: George Allen and Unwin.

Jenkins, T., Zoutendyk, A. and Steinberg, A.G. (1970). Gammaglobulin groups (Gm and Inv) of various southern African populations. *American Journal of Physical Anthropology*, 32: 197–218.

Jerison, H.J. (1973). *Evolution of the Brain and Intelligence.* New York, London: Academic Press.

Kark, S.L. (1954). Patterns of Health and Nutrition in South African Bantu. M.D. Thesis, University of the Witwatersrand, Johannesburg.

Keith, A. (1948). *A New Theory of Human Evolution.* London: Watts.

Keith, A. (1950). *An Autobiography.* London: Watts.

Kenntner, G. (1963). Veränderungen der Körpergrösse des Menschen. Ph.D. Thesis, Universität des Saarlandes, Karlsruhe.

Koenigswald, G.H.R. von (1956). *Meeting Prehistoric Man.* London: Thames and Hudson.

Leakey, L.S.B. (1937). *White African.* London: Hodder and Stoughton.

Leakey, L.S.B. (1953). *Adam's Ancestors*, revised ed. New York: Harper.

Leakey, L.S.B. (1974). *By the Evidence: Memoirs, 1932–1951.* New York and London: Harcourt Brace Jovanovich.

Leakey, L.S.B., Tobias, P.V. and Napier, J.R. (1964). A new species of the genus *Homo* from Olduvai Gorge. *Nature*, London, 202: 7–9.

Leakey, M.D. (1971). *Olduvai Gorge.* Vol. III. *Excavations in Beds I and II, 1960.* London: Cambridge University Press.

Leakey, M.D. (1984). *Disclosing the Past: An Autobiography.* London: Weidenfeld and Nicolson.

Lowe, C. van Riet (1938). The Makapan Caves: an archaeological note. *The South African Journal of Science*, 35: 371–381.

Makino, S. (1951). *An Atlas of the Chromosome Numbers in Animals.* Iowa: State College University Press.

Mann, A.E. (1975). Palaeodemographic aspects of the South African Australopithecines. *University of Pennsylvania, Publications in Anthropology*, 1: 1–171.

Matthey, R. (1949). *Les Chromosomes des Vertébrés.* Lucerne: F. Rouge.

Mayr, E. (1942). *Systematics and the Origin of Species.* New York: Columbia University Press.

Mayr, E. (1970). *Populations, Species and Evolution.* Cambridge, Mass.: Harvard University Press.

Mead, M., Dobzhansky, T., Tobach, E. and Light, R.E., eds. (1968). *Science and the Concept of Race*. New York and London: Columbia University Press.
Montagu, A. (1972). *Statement on Race*, 3rd ed. London: Oxford University Press.
Morell, V. (1995). *Ancestral Passions: The Leakey Family and the Quest for Humankind's Beginnings*. New York, London, Toronto, etc.: Simon and Schuster.
Morris, A.G. (1992). *The Skeletons of Contact: A Study of Protohistoric Burials from the Lower Orange River Valley, South Africa*. Johannesburg: Witwatersrand University Press.
Needham, J. (1942). *Biochemistry and Morphogenesis*. Cambridge: Cambridge University Press.
Needham, J. (1943). *Time the Refreshing River*. London: George Allen and Unwin.
Nurse, G.T. and Jenkins, T. (1977). *Health and the Hunter-Gatherer: Biomedical Studies on the Hunting and Gathering Populations of Southern Africa*. Basel: S. Karger.
Nurse, G.T., Weiner, J.S. and Jenkins, T. (1985). *The Peoples of Southern Africa and their Affinities*. Oxford: Clarendon Press.
O'Dowd, M.C. and Didcott, J.M. (1954). *The African in the Universities*, 2nd ed. Cape Town: National Union of South African Students.
Omoto, K. and Tobias, P.V., eds. (1998). *The Origins and Past of Modern Humans – Towards Reconciliation*. Singapore: World Scientific Publishing Co.
Painter, T.S. (1925). Chromosome numbers in animals. *Science*, 61: 423.
Raikes, R. (1967). *Water, Weather and Prehistory*. London: John Baker.
Robinson, J.T. (1954). The genera and species of the Australopithecinae. *American Journal of Physical Anthropology*, 12: 181–200.
Robinson, J.T. (1956). *The Dentition of the Australopithecinae*. Pretoria: *Transvaal Museum Memoir No. 9*.
Robinson, J.T. (1972). *Early Hominid Posture and Locomotion*. Chicago: University of Chicago Press.
Salzano, F.M., ed. (1975). *The Role of Natural Selection in Human Evolution*. Amsterdam: North Holland.
Schapera, I. (1930). *The Khoisan Peoples of South Africa: Bushmen and Hottentots*. London: Routledge.
Schapera, I. (1939). A survey of the Bushman question. *Race Relations*, Vol. VI: 68–83.
Schultze-Jena, L. (1928). Zur Kenntnis des Körpers der Hottentotten und Buschmänner. *Jenaische Denkschriften*, 17: 147–228.
Simpson, G.G. (1945). The principles of classification and a classification of mammals. *Bulletin of the American Museum of Natural History*, 85: 1–350.
Simpson, G.G. (1953). *The Major Features of Evolution*. New York: Columbia University Press.
Singer, R. and Weiner, J.S. (1963). Biological aspects of some indigenous African populations. *Southwestern Journal of Anthropology*, 19: 168–176.
Steiner, G. (1998). *Errata: An Examined Life*. London: Phoenix (Orion Books).
Stent, G.S. (1972). Prematurity and uniqueness in scientific discovery. *Scientific American*, 227: 84–93.
Tanner, J.M. (1962). *Growth at Adolescence*, 2nd ed. Oxford: Blackwell.

Tobias, P.V. (1966). The peoples of Africa south of the Sahara. In: P.T. Baker and J.S. Weiner, eds. *The Biology of Human Adaptability*, pp. 111–200. Oxford: Clarendon Press.

Tobias, P.V. (1967). *Olduvai Gorge*. Vol. 2: *The Cranium and Maxillary Dentition of* Australopithecus (Zinjanthropus) boisei. Cambridge: Cambridge University Press.

Tobias, P.V. (1969). *Freedom and the Universities*. Johannesburg: University of the Witwatersrand.

Tobias, P.V. (1971). *The Brain in Hominid Evolution*. New York: Columbia University Press.

Tobias, P.V. (1975). Long or short hominid phylogenies? Paleontological and molecular evidences. In: F.M. Salzano, ed. *The Role of Natural Selection in Human Evolution*, pp. 89–118. Amsterdam: North Holland.

Tobias, P.V. (1975). Stature and secular trend among South African Negroes and San (Bushmen). *South African Journal of Medical Sciences*, 40: 145–164.

Tobias, P.V. (1975). Anthropometry among disadvantaged peoples: studies in southern Africa. In: E.S. Watts, F.E. Johnston and G.W. Lasker, eds. *Biosocial Interrelations in Population Adaptation, World Anthropology Series*, pp. 287–305. The Hague: Mouton.

Tobias, P.V. (1976). *The Sixth Freedom*. The Edgar Brookes Academic and Human Freedom Lecture, 1976. Pietermaritzburg: Students' Representative Council of the University of Natal.

Tobias, P.V., ed. and contrib. (1978). *The Bushmen: San Hunters and Herders of South Africa*. Cape Town: Human and Rousseau.

Tobias, P.V. (1978). A little known chapter in the life of Eduardo Mondlane. *Genève-Afrique*, 16: 119–124.

Tobias, P.V. (1980). L'évolution du cerveau humain. *La Recherche*, 11: 282–292.

Tobias, P.V. (1981). *The Evolution of the Human Brain, Intellect and Spirit*. Adelaide, Australia: University of Adelaide.

Tobias, P.V. (1981). The emergence of man in Africa and beyond. *Philosophical Transactions of the Royal Society of London*, B292: 43–66.

Tobias, P.V. (1982). *Man the Tottering Biped: The Evolution of his Posture, Poise and Skill.* Sydney, Australia: Committee in Postgraduate Medical Education, The University of New South Wales.

Tobias, P.V. (1983). Apartheid and medical education: the training of black doctors in South Africa. *The International Journal of Health Services*, 13: 131–153.

Tobias, P.V. (1984). *Dart, Taung and the 'Missing Link'*. Johannesburg: Witwatersrand University Press.

Tobias, P.V. (1985). History of physical anthropology in southern Africa. *Yearbook of Physical Anthropology*, 28: 1–52.

Tobias, P.V., ed. and contrib. (1985). *Hominid Evolution: Past, Present and Future*. New York: Allan R. Liss, Inc.

Tobias, P.V. (1987). The brain of *Homo habilis*: a new level of organization in cerebral evolution. *Journal of Human Evolution*, 16: 741–761.

Tobias, P.V. (1991). *Olduvai Gorge*. Vols. 4A and 4B. *The Skulls, Endocasts and Teeth of* Homo habilis. Cambridge: Cambridge University Press.

Tobias, P.V. (1991). *Images of Humanity: Selected Writings of Phillip V. Tobias*. Johannesburg: Ashanti Publishing.

Tobias, P.V. (1992). *La Paléoanthropologie*. Paris: Editions Mentha.

Tobias, P.V. (1992). *Il Bipede Barcollante*. Torino, Italy: Giulio Einaudi Editors s.p.a.

Tobias, P.V., Raath, M., Moggi-Cecchi, J., and Doyle, G.A., eds. (2001). *Humanity from African Naissance to Coming Millennia*. Florence: Firenze University Press and Johannesburg: Witwatersrand University Press.

Tuttle, R.H., ed. (1975). *Palaeoanthropology, Morphology and Palaeoecology*. The Hague: Mouton.

Vogel, J., ed. and contrib. (1984). *Late Cainozoic Palaeoclimates of the Southern Hemisphere*. Rotterdam and Boston: A.A. Balkema.

Waddington, C.H. (1939). *An Introduction to Modern Genetics*. London: Allen and Unwin.

Washburn, S.L. (1951). The new physical anthropology. *Transactions of the New York Academy of Science*, Ser. II, 13: 298–304.

Washburn, S.L., ed. (1963). *Classification and Human Evolution*. Chicago: Viking Fund Publications in Anthropology.

Watts, E.S., Johnson, F.E. and Lasker, G.W., eds. (1975). *Biosocial Interrelations in Population Adaptation, World Anthropology Series*. The Hague: Mouton.

Weidenreich, F. (1936). The mandibles of *Sinanthropus pekinensis*: a comparative study. *Palaeontologia Sinica*, Ser. D. 7: 1–162.

Weidenreich, F. (1943). The skull of *Sinanthropus pekinensis*: a comparative study of a primitive hominid. *Palaeontologia Sinica*, 10: 1–298.

Wells, L.H. (1937). The status of the Bushman as revealed by a study of endocranial casts. *South African Journal of Science*, 34: 365–398.

Wells, L.H. (1960). Bushman and Hottentot statures: a review of the evidence. *South African Journal of Science*, 56: 277–281.

White, M.J.D. (1945). *Animal Cytology and Evolution*. Cambridge: Cambridge University Press.

White, T.D. (1977). New fossil hominids from Laetoli, Tanzania. *American Journal of Physical Anthropology*, 46: 197–230.

Wolanski, N. (1966). The secular trend; micro-evolution, physiological adaptation and migration, and their causative factors. *Problems of World Nutrition*. Proceedings of the 7th International Congress of Nutrition, Hamburg.

Wolpoff, M.H. (1980). *Palaeoanthropology*. New York: Knopf.

Index

Aberle, Dave 102
abnormalities, congenital (teratology) 23
Aborigines 106
academic apartheid 188
see also apartheid
Académie des Sciences 35
Accademia Nazionale dei Lincei 222
Adler, Cyril and Esther 147
Adler Museum of the History of Medicine 147, 148
Adventures with the Missing Link (R.A. Dart and D. Craig) 237
Afalou-bou-Rhummel skull fragments (Algeria) 86
African Medical Scholarships Trust Fund 60, 200
African Middle Stone Age 46
Alaska (skeletal remains) 137
Allan, John Cameron ('Jack') 120
Allison, Anthony 46
American Association of Physical Anthropologists (AAPA) 102, 124, 135
American Journal of Human Evolution 138
American Journal of Physical Anthropology 136, 138
American Museum of Natural History 113, 228, 229, 233, 234
American Philosophical Society (Philadelphia) 141
Anatomical congress in Milwaukee 128, 129
Anatomical Society of Great Britain and Ireland 34, 102
Anatomical Society of Southern Africa 179
Anatomy Act, The 131
anatomy, P.V. Tobias appointed as lecturer in 63
Ancestors Exhibition 228, 235
Angel, Larry 102
Ann Arbor 118–23
anthropology 94, 95
biological and cultural (synthesis between) 142
branches of 113
historical aspects of 141
palaeo- 32, 42, 101, 157, 178, 179, 220
physical 18, 101, 112, 113, 138, 152, 157, 178, 208
physiological 152
racism and 208
Anthropology Museum, University of Pennsylvania 142
anthropometry 63, 76
anti-apartheid academic revolt 179
anti-apartheid demonstration 235
apartheid 8, 36, 76, 96, 164, 179, 181, 223
boycotts 194–96
era 193
in the university arena 184–92
ape family (Pongidae) 215
ape-man 25
see also australopithecine
applied embryology *see* embryology, applied
Arambourg, Camille 83
Arcy-sur-Cure (Cro-Magnon type) 83
Arnold, Maurice ('Toby') 115, 120
Asselar skull (Mali) 86
asthma 65 66, 71
Atlantic Charter, Four Freedoms of the 56
Atlas of Men (W. Sheldon) 205, 206
atomic bomb (Hiroshima) 34
Australian and Tasmanian people 107
australopithecine 95
adult ape-man 52
deposits 43
evolutionary development of 126
lecture on 124
(South African ape-man) site 25

Australopithecus 37, 40, 41, 43, 88, 89, 90, 215, 216
africanus 127, 215, 218, 219
boisei 243
pattern of mosaicism in 216
prometheus 41
robustus 127
Ayala, Francisco José 222, 223, 225

Baartman, Sarah (Saartjie) 84–86
baboon
Chacma 33
crania, mandibles and endocasts 51
fossil, at Makapansgat 40
fossil, at Taung 37
giant 52
and monkey skulls 37, 43
Baker, Burton 129
Balinsky, Boris 151
Balsan, Francois 64, 65, 66, 67, 68
Baltimore 99
Barbour, George and Dorothy 130
Barkham, Percy 16, 71
Barlow, George 37, 50, 51
Bartelmez, George W. 99
Bechuanaland 168
Becker, B.J.P. ('Bunny') 178
behaviour, genetics of 222
Berkeley 100, 158
Berlin Museum 247
Biochemistry and Morphogenesis (J. Needham) 32
biodiversity 17
Biological Basis of Human Freedom, The (T. Dobzhansky) 225
biological basis of language 102
Biology of Ultimate Concern, The (T. Dobzhansky) 225, 226
biometrical approach 138
Biometric School 70, 71, 77
bipedalism 27, 28
Birdsell, Joseph B. 106, 107
Birmingham Medical School in Alabama 129
Black Africans in Union of South Africa 8
Black, Davidson 95, 231, 237
Blacking, John 160
Black Sash 235
Bloemfontein 8, 9, 11
Blombos Cave 48
Bobrow, Martin 160
Bolt's Farm 31
Bone Age 216
Boné, Edouard 153
'Bone-Picker, Mr' 26
Bone, Roger C. 9, 155
Border Cave 48
Boshier, Adrian 48
Boston 144
Botswana 69, 145
Bouman, K.H. 230
boycotts during apartheid 194
Boyes, J. Wallace 151
Brain, Charles Kimberlin ('Bob') 40, 217, 227
breccia 25, 38, 40, 42
Brenner, Sydney 1, 29, 33, 61, 71, 73, 103, 178
Breuil, Abbé Henri 153
British Academy 194
Britten, 'Dassie' 10
Broken Hill 81
see also Kabwe
Bronkhorstspruit 31
Bronowski, Jacob 35
Broom, Robert 1, 37, 39, 40, 41, 43, 51, 52, 53, 88, 212, 215, 237
Brown, Dolfanna and Lester 9
Burkitt, Miles 75, 157
Burwitz, Nils 174
Bushmen of the Kalahari *see* San (Bushmen) of the Kalahari
Buxton Limeworks at Taung 214, 237

Calder, Ritchie 35
calotte, Javanese 230
Cambridge (Massachusetts) 144
Cambridge University 34, 59, 70–75, 76, 80, 88

Anatomy Department 92, 93
Faculty of Archaeology and Ethnology 78
last months in 92–97
see also Duckworth Laboratory
Campbell Collection 17
Campbell, Revd John 86
Campbell, Roy 14
Carnegie Institution 99
Carter, Lindsay 204
Casablanca (fossil, Morocco) 242
Caspersson, T. (Karolinska Institutet, Stockholm) 151
Catholic studies of fossil hominids 153
Catholic University of Louvain 153
Cave of Hearths 38
Cecchi, Jacopo M. 1
cell division
DNA threads in metaphase of 29
premeiotic mitosis 33
cercopithecids (baboons and monkeys) 37, 39, 40, 41, 51
Cercopithecoides williamsi 40
cerebrotonia 208
Champion, Peter 83
Charles Sumner lectureship 134
Chemical Embryology (J. Needham) 32
Chicago 124
Chicago Museum of Industry and Science 126
Chicago Natural History Museum 127
Chinese and Javanese 'prehominids' 231
Choma, Zambia 161
chromosomes 32, 178
and sex cells of the gerbil 70
during cell division 29
mammalian 29, 151
of human and other mammals 32
of the elephant shrew 29
study of (cytogenetics) 11, 16, 24
Chromosomes, Sex-Cells and Evolution in a Mammal (P.V. Tobias) 80, 94
Chubb, E.C. 17
Cincinnati, Ohio 130
Clark, Desmond and Betty 157, 158, 159, 160
Clark, Wilfrid Edward LeGros 79, 94, 219
Clarke, Ronald John 27
Cleveland, Ohio 131
cloning of individuals 36
Cluver, Michael 234
Cobb, W. Montague ('Monty') 117, 133, 134, 135
Cold Spring Harbor 151–52
colobine monkey 41
Colson, Elizabeth 159
Columbia University 153, 204, 221
Combe Capelle, remains 246, 247, 248
Coming of Man: Was it Accident or Design? (R. Broom) 215
Conroy, Glenn 116
Coon, Carlton S. 125, 142, 143
Copley, Bruce 209
Correnti, Venerando 239
Coryndon, Shirley 233
Council for Scientific and Industrial Research *see* CSIR
Cowan, Maxwell 103
Crafts, Professor 130
Creation Science 228, 229
Cro-Magnon (fossil hominids) 83
CSIR (Council for Scientific and Industrial Research) 80, 235
Cushing, Harvey 146
Cuvier, Georges 85
cyclothymic personality 207
cytogenetical research 33, 151
cytogenetics 11, 32, 94, 95, 101, 178
cytoplasmic hypothesis 111

Damon, Al 166
Dart, Diana and Galen 26, 27
Dart, Dora 213
Dart, Raymond Arthur 1, 24–27, 30, 37, 38, 40–43, 50, 51, 53, 63, 68, 70, 78, 88, 89, 92, 95, 106, 107, 111, 114, 126, 131, 135, 153, 164, 174, 175, 176, 177, 202, 211, 212, 213, 214, 215, 216, 217, 218, 219, 221, 223, 234, 237, 238

Darwin, Charles 141, 215, 221, 226, 227
Davis, Dwight 127
Day, Michael 229
de Chardin, Teilhard 153, 230
De Humani Corporis Fabrica (Vesalius, 1543) 147
Delaney, Brendan 108
de l'Estrange, Monique 83
Delson, Eric 228, 229, 233, 234
Dempsey, Ed 111, 112
deoxyribose nucleotides 151
Descent of Man, The (C.R. Darwin) 215, 223
Detroit 124
Dinath, Yusuf 116, 191
Dinopithecus ingens 52
Dirè-Dawa mandibular fragment (Ethiopia) 86
Dixon, Roland 101
Djebel Kafzeh cranium (Israel) 86
DNA
  discovery of 196
  structure of 68, 122
  threads in metaphase of cell division 29
Dobzhansky, Theodosius ('Doby') 32, 70, 221, 222, 223, 224, 225, 226, 227
Doctor of Philosophy degree 70
Doherty, J.G. 153, 246
Dreyer, Thomas F. 238
Drosophila 221, 223, 227
Dubois, Eugene 230, 231
Duckworth, Wynfrid Laurence Henry 92–94
Duckworth Laboratory (Cambridge) 71, 76–82, 92, 93, 232
  see also Tobias, Phillip Vallentine, Nuffield Senior Travelling Fellowship (Cambridge)
Dunn, L.C. 221
Du Plessis, D.J. ('Sonny') 177, 180
Durban 9, 13
Durban High School 11, 12, 14
Durban Natural History Museum 16, 17
Duruthy (Cro-Magnon type) 83
ecology 23
  of Kalahari San (Bushmen) 24
ectomorphic (people) 204, 209
Edgar Brookes Academic and Human Freedom Lecture (1976) 199
Education League of South Africa 133
Edward Philip Stibbe Postgraduate Laboratory 213, 214
'Egbert' (Ksâr 'Akil) 246
Eiseley, Loren C. 141
elephant shrew 29, 33
embryology 178
  applied 23
  descriptive and experimental 101
  human 23
  of the sex glands 100
embryonic liver cells 99
endomorphic (people) 204, 209
Erikson, Hertha (later de Villiers) 149
Errata: An Examined Life (G. Steiner) 174
Eskimos *see* Inuit
ethnic or tribal universities in South Africa 198
ethnomusicology 160
euchromatin 151
evolution
  Catholic attitude to 154
  genetics and 31
  human, synthetic approach to 152
  of language 142
  of the brain and spoken language 155
  of the horse family 153
evolutionary scale 143
Evolution, Genetics and Man (T. Dobzhansky) 225
Ewing, Father Franklin (Fordham University) 153, 246
Ezekiel, Prophet, Book of 17–18

Fagan, Brian 160
Farini, Gilarmi A. ('Farini the Great') 67
  see also Hunt, William Leonard
Farrell, Brian 30

Federative International Committee on
Anatomical Terminology 87
Fellowship of the College of Surgeons 177
Fifth Freedom (of the Academy) 197, 198, 199
Findlay, George 52
First International Congress of Human Palaeontology 229
Fitzpatrick, James Percy 45
Florisbad cranium (Free State, South Africa) 238
Fontéchevade cranial fragments (France) 86
Ford Center for the Study of the Behavioural Sciences 103
Ford Foundation 103
Fordham University (New York City) 153, 246
'Ford Think Tank' 103
Fort Hare University 57, 58
Fosbrook, H.A. 159
fossil(s)
- algae 108
- ash 42
- baboons and monkeys 25, 40, 43
- baboons from Taung 37
- baboon skull 214
- bones and teeth 178
- casts obtainable from 243
- East African and Asian 232
- hominid remains 80, 81, 228
- hominid repatriation 245
- hominids from China and Java 229, 230
- hominids from Java and East Africa 230
- hominid sites, South African 94
- hominids (Catholic studies) 153
- hominid specimens 83, 179, 230, 231
- repatriation 245
- Saccopastore (Italy) 238
- San Felice Circeo (Monte Circeo, Italy) 238

fossils from
- Chad 242
- East Africa 232
- East Africa and Asia 232
- Java 230
- Mount Carmel 242
- Northern Kenya 242

Fourth World (hunters and foragers) 167
Free Circulation of Scientists 193
Freedman, Rabbi A.H. 19
Freedoms of the Academy 197, 198, 199, 201
Frelimo, Mozambique 62
French Panhard-Capricorn Expedition *see* Panhard-Capricorn Expedition
Fulton, John 146

Gadjah Mada University (Indonesia) 245
Galloway, Alexander 39, 50, 51
Garankuwa, Medical University 199
Gardner, William 146
Gautscha Pan in South West Africa (Namibia) 145
Geffen, Lawrence 160, 163
gene frequencies of populations 152
general election (1948) 8
genetic
- causation of morphological traits 125
- code 122, 196
- engineering 36, 196

Genetic Diversity and Human Equality (T. Dobzhansky) 225
geneticists, human 121
genetics 16, 24, 178
genetics and evolution 31, 32
Genetics and the Origin of Species (T. Dobzhansky) 32, 221, 227
genetics of behaviour 222
gerbil, chromosomes and sex cells of (Ph.D. thesis) 70, 80
Giddings, J. Louis 142
Gigantopithecus 241
Gilbert, Christine 23
Gillett, Rhonda M. 161
Gillman, Joseph 23, 29, 30, 100, 219
Gillman, Theodore 175
Glauber, Dennis 25
Glossina morsitans *see* tsetse fly

Goldschmidt, Richard 32
Goodenough, Ward 142
Gough, Kathleen 102
Grahamstown 66
Grantchester 73, 74
Greathead, Val 218
Great Synagogue, Johannesburg 20
Greulich, William 103, 146
Growth and Adolescence (J. Tanner) 167
Gruber, Jacob 141
Gunther, H.F.K. 77, 209
Gwembe Valley Survey 159, 160, 161, 205

habilis 218
Haldane, J.B.S. 76
Hamman, Carl August 113
Hamman-Todd Collection (human skeletons) 114
Hankey (Eastern Cape) 86
Hauser, Otto 86, 247
Health and the Hunter-Gatherer (G. Nurse and T. Jenkins) 153
Henneberg, Maciej 182
Henshilwood, Chris 48
heredity, chromosomes and 16
heredity counselling 12, 122
Hertig-Roc embryo 99, 100
Hiroshima 34, 36, 121
Historical Monuments Commission 31, 53, 54
Historic Cave 38
History of Man (C.S. Coon) 142
History of Medicine Society 147
Hitler, Adolf 238
Hitler regime 77
Hoffman, A.C. 238, 248
Hofmeyr, Jan Hendrik 56, 58, 197
  Fifth Freedom 197
Holliday, Clayton 160
Holloway Commission 188
Holocene, the 143
hominid
  5 Olduvai 82
  Australopithecus 215
  early, South African 219
  evolution 81, 216
  fossils *see* fossil(s), hominid
  jawbone from Sidi Abderrahman 83–84
  non-, activities 217
  Olduvai 219
  specimens 37, 40
Hominidae 215
Homo erectus 143, 231, 240, 241
Homo habilis 78, 142, 218, 219, 231
Homo kanamensis 81
Homo sapiens 143
Homo sapiens sapiens 81
Homo-Simiadae (Man-Ape Family) 215
Honeyman-Heath, Barbara 204
Hooton, Earnest Albert 101, 106, 107, 137, 209
Howard University Medical School 134
Howell, Francis Clark 124
Howells, William W. 220
Hrdlicka, Ales 101, 136, 137, 237
Hughes, Alun 43, 235, 236
human embryology *see* embryology, human
Human Genetics Conference (Wits, 1962) 151
human
  evolution 179
  evolution, synthetic approach to 152
  geneticists 121, 122
  genome project 36
  heredity clinics 120
  morphology 214
  palaeontology 212, 233
    see also palaeo-anthropology
  physical tendencies 204
  secular trend in 166–68
Hunt, William Leonard ('Farini the Great') 67
Hunterian Museum 177, 213
hunters and foragers (Fourth World) 167
hyena, extinct species of 53

insect cyto-taxonomy and cyto-phylogeny 151

Inskeep, Ray 160
Institut de Paléontologie Humaine 86, 242
Institute for the Study of Human Variation 152
Institute for the Study of Mankind in Africa 223, 225
Institute of Vertebrate Palaeontology and Palaeo-anthropology 246
International Anatomical Nomenclature Committee 87
International Council of Scientific Unions (ICSU) 193, 196
International Union of Prehistoric and Protohistoric Sciences (IUPPS) 194
Inuit (Eskimos) 142
Irvine, Lee 14
Izod, Edwin Gilbert and 'Pat' 37, 214

Jacob, Teuku 241, 245
Jacobs, Leopold 205
Java Man 230, 240
Javanese fossils 232, 241
Jebel Irgoud (fossil, Morocco) 242
Jefferson Medical College (Philadelphia) 102
Jeffreys, M.D.W. 203
Jenkins, Trefor 122, 151, 153, 165–66, 225
Jensen, Joseph Stokes 37, 40
Jewish education 19
Jock of the Bushveld (J.P. Fitzpatrick) 45, 46
Jones, Owen 46
Jones, Trevor 37, 43, 50, 52
Journal of Human Evolution 169
Judaism *see* Jewish education

Kabwe Man 244
Kabwe or Broken Hill cranium 125, 242
Kabwe (Zambia) 81
Kalahari 66
  Bushmen 63
  dwellers, prehistory 68
  French Panhard-Capricorn Expedition 63–68
    see also Panhard-Capricorn Expedition
  'Lost City' 64
    see also 'Lost City' of the Kalahari
  Research Committee 69, 145
  San Bushmen 83
Kanam jaw fragment (Kenya) 81
Kanjera cranial pieces (Kenya) 81
Kappers, C.U. Ariëns 230
Kariba Dam, Tonga people and the 157–64
Kark, Sidney L. and Emily 167
Keith, Arthur 79, 95, 101, 138, 216, 217
Kenntner, Georg 169
Kenya (fossil hominid specimens) 81
Keppel-Jones, Arthur 176
Kgalagadi Transfrontier Park 67
Khoe-Khoe 68
Khoe-San 83, 84, 85, 94, 95, 107, 164
King, G.B. 17
Kitching, James W. 37, 40
Klomfass, Roland 177
Klug, Aaron 14
Koenigswald, G.H.Ralph von 1, 229, 230, 231, 232, 233, 239, 240, 241, 245
Koller, P.C. 33
Kramer, Beverly 179
Kretschmer, Ernst 207
Kromdraai 31, 39, 53, 54
Ksâr 'Akil (cave deposit, Lebanon) 153, 246

La-Chapelle-aux-Saints (France) 83
Ladybrand, Rose Cottage Cave 31
Laetoli (Tanzania) 229
La Ferrassie (France) 83, 86
La Madeleine (Cro-Magnon type, France) 83
Lane, Tony 151
language biological basis of 102
  evolution of 142
La Quina (France) 83
Lascaux (subterranean art gallery) 86, 87
Later Pleistocene 143
Later Stone Age 48, 84
  in the Kalahari 67
Lawrence, T.W.P. 81

Leakey, Louis 1, 40, 41, 76, 78, 81, 82, 218, 219, 231, 232, 243
Leakey, Jonathan 218
Leakey, Mary 1, 82, 218, 231, 232, 243, 246
Leakey, Richard 244, 246
Lee-Thorpe, Julie 127
Le Moustier Cave (France) 247
Le Moustier (juvenile Neandertal skeleton) 86, 246, 247, 248
Le Riche, William Harding 37, 51, 52, 67
Levine, Errol 178
L'expédition Panhard-Capricorne (F. Balsan) 66
Library of the History of Medicine (Yale) 146
Limeworks, Makapansgat 25
  see also Makapansgat
'Little Foot' 27
liver cells, embryonic 99
Living Anatomy 115
Lombaard, B. 54
London, Vallentines 19
L'Origine dell 'Uomo 223
Los Angeles and the Grand Canyon 105–10
'Lost City' of the Kalahari 64, 66, 67
Louw, Gert 67
Ludwig, Edeltraud 169
lumping 41, 42

MacCrone, I.D. 176
Maguire, Brian 45
Maguire, Constantine Duncan 47
Makapansgat 37–44, 45, 49, 51, 216
  bones 217
  Limeworks 25, 31, 89, 153
Makino, Sajito 33
Malan, 'Berry' 31, 46, 78
Malan, Daniel Francois 184
Malarnaud (France) 83
malocclusion, postural aspect of 25
mammalian chromosomes 29
Man-Eaters of Tsavo 128
Mangera, Dawood 160
Mankind Evolving (T. Dobzhansky) 225
Manouvrier, Léonce-Pierre 136
Man's Anatomy (P.V. Tobias and M. Arnold) 120
Manzi, Giorgio 238
Marquard, Leo 56
Marshall, Lawrence and Lorna 144, 145
Martin, Adrian R.P. 160
Matthey, Robert 33
Maubane, Eriam 228
Maytom, Peter H.B. 14, 22
McCown, Theodore 100, 102
McKee, Jeff 220
Mead, Margaret 204
Medical Association of South Africa 214
Medical-Chirurgical Society of the District of Columbia 134
Medical History Institute, Leipzig 146
Medical University of South Africa (Medunsa) 192
Meiring, A.J.D. 238
Melanesia 142, 204
mesomorphic (people) 204, 205
mesomorphs 209
Middle Pleistocene 143
Middle Stone Age
  deposits 47, 48
  implements 42
  in the Kalahari 67
Millin, Sarah Gertrude 120
Minugh-Purvis, Nancy 246
Mogg, Albert O.D. 17
Mojokerto child fossil (Java) 245
Mollet, Ollerais 40, 45
Mollison, Theya 80
Mondlane, Eduardo Chivambo 61, 62
Montague, Ashley 170
Montja, Peter 228
Mordecai, Ben and Zema 176
Morocco (fossils from) 242
morphogenesis 32
morphological features 107
morphological traits and genetic causation 125
morphology, human 214
Morris, Roy 205

Mount Carmel fossils 242, 243
Mourant, Geoffrey 76
Musée de l'Homme (Paris) 83, 113, 242
Museum of the University of Pennsylvania 141
Muséum National d'Histoire Naturelle (Paris) 83, 84, 242
Mwulu's Cave 45–48

nagana *see* tsetse fly
Nagasaki 36, 121
Napier, John 218, 219, 231
National Geographic 235
National Institutes of Health (NIH), Bethesda 139
National Monuments Council *see* South African National Monuments Council
National Museum in Bloemfontein 238
National Museum in Dar-es-Salaam 244
National Museum in Nairobi 243
National Party 57, 59, 96, 184
National Party's education policy 57
National Union of South African Students 56, 58, 198, 200
  see also Nusas
Natural History Museum, Frankfurt 241
Natural History Museum, London 73, 80, 83, 93, 113, 242, 244
natural selection (Charles Darwin) 32
Natural Superiority of Women, The (A. Montague) 170
Navaho Indians 102
Nazism 36
Neandertaloid human remains 153
Neandertaloid (or Rhodesioid group) 125
Neandertal skulls (Italian) 238
Neandertal specimens in Musée de la Homme, Paris 83
Needham, Joseph 32
Neel, James V. ('Jim') 12, 120, 122
  neuro-anatomy 101
New Britain (Melanesia) 142
New York City 94, 98, 99, 149–50, 204, 208, 222, 229, 233, 246
Ngandong skulls (Java) 245, 248
Ngwenya (Swaziland) 48
non-hominid activities 217
Norden, Albert Louis ('Bert', P.V. Tobias's godfather and later stepfather) 5, 7, 12, 16, 18, 22, 34, 176
Norden, Fanny (P.V. Tobias's mother) 16, 22, 34, 176, 220
  see also Rosendorff, Fanny and Tobias, Fanny
Northern Rhodesia *see* Zambia
Norwich, I. ('Oscar') 147
nuclear fixative 111
nucleic acid cycles 151
nucleolar organisers 128
nucleoli, complement of 99
Nuffield Foundation 69, 70
Nuffield Senior Post-Doctoral Fellow 34
Nurse, George T. 153
Nusas 33, 56–60, 198, 199, 200, 201
  P.V. Tobias as president of 58, 59, 62, 65, 197, 200
Nyerere, President Julius 244

Oakley, Kenneth 79, 80
occipital bone, curvature of 79
ochre 46, 47, 48
Oguma, Kan 33
Oldowan stone tools 218
Old Stone Age implements 42
Olduvai (Tanzania) 242
  fossil hominids 231
  Gorge 218, 229, 243
  hominid 5, 82, 218
  specimens 219
Old Yellow South Africans, The 95
Ooikolk 67
Open Universities 62
Open Universities of South Africa 198
Oppenheimer, J. Robert 34
Origin of Races, The (C.S. Coon) 143
Origin of Species, The (C.R. Darwin) 31, 221
Oschinsky, Lawrence 76
osteodontokeratic culture 43, 126, 216

Painter, Theophilus 33
palaeo-anthropology 32, 42, 101, 157, 178, 179, 220, 231
palaeontologists 27, 212
paleontology, human 121, 212, 233
Palaeontology, Human, First International Congress of 229
Palo Alto 102
Pan-African Congress of Prehistory and Palaeontology (1959) 81
Pan-African Congress on Prehistory (1955) 142
Panhard-Capricorn Expedition 65, 67, 68, 69, 84, 145
Papio darti 38
Paranthropus 41
Parapapio broomi 43
Parapapio major 39
Passarello, P. 239
Paterson, Brian 128
Paterson, Hugh 224
Paton, Alan 213
Patten, Bradley and Barbara 118, 132
Peabody Museum (Harvard) 145, 242
Peking Man, skull 95, 229, 230, 237, 240, 241, 245
Peninj mandible (Tanzania) 229, 231
Peoples of Southern Africa and their Affinities, The (G.T. Nurse, J.S. Weiner and T. Jenkins) 153
personality and behaviour, three psycho-components 208
Peterson, Roy 116
Phillips, John 23, 24, 38
physical anthropology 93, 112, 157, 178, 208
physical appearance and personality 207
physique, study of 203, 204, 205, 206, 208
physiques, sports 209
Pithecanthropus finds 230, 231
Plesianthropus ('Mrs Ples') 41, 54
political awakening 55–62
Pongidae (ape family) 215
Pontifical Academy of Sciences 155
Pope John Paul II 155
population genetical studies 122
Posner, Dora 9, 12
Posner, Harry 9, 12
Posner, Hyam 9, 12
Posner, June Elfrieda 9
Posner, Walter 9, 12
posture, erect, evolution of 25
Powers, Rosemary 80
Prehistoric Cultures of the Horn of Africa, The (J.D. Clark) 157
Prehistory of Southern Africa, The (J.D. Clark) 157
Price, Max 147
premeiotic mitosis 34, 95

Rabat (Morocco) 86
race and the races of living humanity 76
Race Classification Act 69
Race Crossing in Man: The Analysis of Metrical Characters (J.C. Trevor) 77
Racial Elements of European History, The (H.F.K. Gunther) 77
racialistic anthropology 208
Rainbow Cave 38, 42
Rand Afrikaans University (RAU) 189
Raymond Dart Collection of Human Skeletons 114
Raymond Dart Lectures 223, 224
Recent Advances in the Evolution of Primates (Vatican) 155
Redfield, Robert 124
Reeves, Bishop Ambrose 96
repatriation of fossils 245, 246
Reynolds, Barry 159
Rhodesian Man 81, 244
Rhodesioid group 125
Rhodes-Livingstone Foundation (Lusaka) 159
Rhodes-Livingstone Museum (Livingstone) 157, 158
Rhodes University 57, 65
Ricklan, David 178
Rijksuniversiteit (Utrecht) 241

Robinson, John 1, 27
rock art, prehistoric 153
Rockefeller Foundation 88, 97, 98, 213, 241
Rockefeller Museum (Jerusalem) 242
Rockefeller University (Institute) 222
Romanes, George 96
Romer, Alfred S. 237
Rooi Pits 67
Rose Cottage Cave, Ladybrand 31, 46
Rosendorff brothers (Max, Siegfried, Herman and Karlie) 12
Rosendorff, Clive 192
Rosendorff, Fanny (P.V. Tobias's mother) 3
see also Norden, Fanny and Tobias, Fanny
Rosendorff, Frieda 11, 16
Rosendorff, Martin 4
Rosendorff, Vincent 22
Royal Anthropological Institute of Great Britain and Ireland 102
Royal Society (London) 14, 35, 194, 212
Royal Swedish Academy of Sciences 244, 245

Saccopastore (fossil, Italy) 238, 239, 248
Salmons, Josephine 37, 214
Sambungmachan skull (Java) 245
San (Bushmen) of the Kalahari 63, 64, 65, 68, 69, 83, 84, 106, 122, 124, 125, 142, 145, 151, 165, 167, 168, 170, 172, 204, 205
see also Khoe-San
San Felice Circeo (fossil, Italy) 238, 239, 248
Sangiran cranium (Java) 230, 240
Sandburg, Carl 201
Santa Maria della Pieta (Rome) 239
Schepers, Gerrit 51, 52
scientists, role of 36
Scudder, Thayer 159
Second World War 14–16, 19
secular trends and changes in humans 166–71
Segre, Aldo 239
Segre-Naldini, Eugenia 239
Senckenberg Research Institute 241
Separate Universities Bill 185
Sergi, Sergio 238, 239
Seven Caves (C.S. Coon) 142
sex glands, embryology of 100
sexual dimorphism 170–72
Shapiro, Harry L. 234
Sheldon, William H. 203, 204, 205, 206, 207, 208, 210
Shellshear, Joseph 213
Sigerist, Henry 146, 147
Silberberg Grotto 53
Silberberg, Helmuth 53
Silver Jubilee Congress of Nusas (1949) 199
Simpson, George Gaylord 227
Sinanthropus and *Pithecanthropus* 231
Singa cranium (Sudan) 242
Sino-Javanese fossils 231
Singer, Ronald 107
Sixth Freedom (of the Academy) 199
skeletal biology 116
Skhūl caves (Israel) 242
skulls, monkey and baboon 37, 39
skulls, primate 39
Smith, G. Elliot 95, 212, 213, 216
Smithsonian Institution 136, 137, 145
Smit, Paul 209
Smuts, Jan Christian 176, 184
Solo skull from Ngandong (Java) 241
somatotonia 208
somatype
components 209
descriptions 205–06
formulae 205
studies 209
somatypes of black people of southern Africa 205
South African Association for the Advancement of Science 238
South African Heritage Resources Agency 229
South African Medical Association 14
South African Medical and Dental Council 214

South African National Monuments Council 229, 234
see also South African Heritage Resources Agency
South Africa's Past in Stone and Paint (M.C. Burkitt) 157
Soviet Genetics Decree (1948) 196
Spencer, Frank 1
splitting 41, 42
St Andrew's School 9, 12
Stanford University 102, 103
Stebbens, Ledyard 223
Steffenson, Jon 76
Steiner, George 174
stem cell research 36
Sterkfontein 31, 39, 43, 49–54
ape-man 53
cercopithecid skulls 37
flora of 17
fossilised bones found at 27, 89
hominid fossils 243
stalactite and stalagmite formations 49, 50
Stewart, T. Dale 101, 117, 135, 136, 137, 138, 140, 152
Stibbe, Edward Philip 213
St Louis 111–17
Stone Age 216
Stone Age implements, Middle 42
Stone Age implements, Old 42
stone tools, Oldowan 218
Storey, Frank 12
Struben, Fred and Harry W. 49
Sudan (Singa cranium) 242
Students' Medical Council (SMC) 212
Students' Representative Council 61, 65, 191
Students' Representative Council, Pietermaritzburg 199
Sullivan, J.W.N. 9
Sutton, W.G. 97

Tainton, Colonel and Doreen 64
Tanganyika (fossils from) *see* Tanzanian fossils
Tanner, James 167, 168
Tanzanian fossils 78, 241, 242
taphonomy 217
Tatera brantsii draco (gerbil) 31
Taung child 37, 42, 43, 89, 211, 215, 216, 220, 228, 234, 235
Taung, fossil baboon from 37
Taung skull 43, 49, 50, 81, 89, 95, 117, 135, 211, 214, 215, 218, 228, 235, 237, 238
Tax, Sol 124
taxonomy 41
teratology 23
Ternifine fossils (Algeria) 242
'Terry Collection' of skeletons 113
Terry, Robert James 113, 213
Theropithecus darti 38
Third World populations (secular changes in) 168
Thomas, Ieuan Glyn 62, 198
Through the Kalahari Desert (G. Farini) 67
Tildesley, Miriam 77
Tobias, Fanny (P.V. Tobias's mother) 3, 4, 5, 6, 7, 11, 12, 16
see also Norden, Fanny and Rosendorff, Fanny
Tobias, Joseph Newman (P.V. Tobias's father) 3, 4, 5, 7, 12, 22, 31, 34, 158, 176
Tobias, Phillip Vallentine
adult life spent in Johannesburg 3
in America 98–156
anti-apartheid campaign in South African universities, launch of 58–59
asthma, onset of 65–66
birth and early childhood on the coast of KwaZulu-Natal 3–7
Dart, Raymond Arthur, influence of 24–26, 211–20
Dean of Medical Faculty, Wits 190
death of sister 11
see also Tobias, Valerie Pearl
Durban during the Second World War 15, 16
Durban, teenage years and influences 16, 19

early childhood events and memories 4, 5
education and studies
Durban High School (matriculated at) 11, 14, 20
Durban Preparatory High School 5
President Brand School, Bloemfontein 9
St Andrew's School, Bloemfontein 9
University of the Witwatersrand, studies at 5, 22, 23, 24, 55
B.Sc. Hons. 22, 31
Medical B.Sc. 22, 30, 56
M.Sc. 32, 34
Ph.D. 11, 33, 34, 70
father, Joseph Newman Tobias 3
see also Tobias, Joseph Newman
in France 83–91
freedoms of the academic world, presentation of 197–99
grandfathers, Phillip Tobias and Martin Rosendorff 3, 4
Health and the Hunter-Gatherer 153
introduction to the field of human anatomy 11
Judaism, introduction to and involvement with 19
Kalahari San or Bushmen, first ecological study of (1951) 23
a lifetime of writing, editing and publishing 5, 9, 10
Living Anatomy classes 115
mother, Fanny Rosendorff 4
see also Norden, Fanny and Tobias, Fanny
negotiator for repatriation of Sarah Baartman's remains 85–91
Norden, Albert Louis (godfather and later stepfather) 7
see also Norden, Albert Louis
Nuffield Senior Post-Doctoral Fellow, Cambridge University 34
Nuffield Senior Travelling Fellowship, Cambridge 70–75
see also Duckworth Laboratory
Nusas, national president 57–58, 197
see also Nusas
opposition to academic apartheid 188
parents' divorce 12
Posner, Dora and Hyam (aunt and uncle) 9, 12
Posner, Harry and Walter (cousins) 9, 12
Posner, June Elfrieda (cousin) 9, 12
publication of book based on Ph.D. thesis 80
'reluctant professor' at Wits Department of Anatomy 174–83
research projects and studies
attainment of the upright posture 27–28
Bolt's farm 31
curvature of the occipital bone 79
Gladysvale 31
human biological study of Tonga people in Zambia 160
Kalahari Research Committee, chairman of 69
see also Kalahari
Kalahari San or Bushmen 23
see also San (Bushmen) of the Kalahari
Kanam jaw fragment and Kanjera cranial pieces 81
Kromdraai 31
see also Kromdraai
Makapansgat Limeworks 31
see also Makapansgat
mammalian chromosomes 29
Mwulu's Cave 45–48
Olduvai specimens analysis of 219, 231
Panhard-Capricorn expedition, member of 65–68
Rose Cottage Cave 31
see also Rose Cottage Cave
San people, short stature 165–73
Sterkfontein 31
see also Sterkfontein

Rosendorff, Fanny (mother) 3
see also Norden, Fanny and Tobias, Fanny
Rosendorff, Martin (grandfather) 4
time spent in Bloemfontein in the 1930s 7, 8
Tobias, Fanny (mother) *see* Norden, Fanny and Rosendorff, Fanny
Tobias, Joseph Newman (father) 3
see also Tobias, Joseph Newman
Tobias, Phillip (grandfather) 3
Tobias, Valerie Pearl (sister) 4
see also Tobias, Valerie Pearl
University of the Witwatersrand *see* Tobias, Phillip Vallentine, education and studies and Wits
Wits lecturer and later head of Department of Anatomy 63, 164, 174–83
Tobias, Valerie Pearl ('Val') 4, 7, 9, 11, 12
Todd, Thomas Wingate 114, 137
Tonga people and the Kariba Dam 157–64
Tonga people of Zambia 23, 122, 160, 161, 204, 205
Traill, Tony 151
Transvaal Museum 51, 113, 145, 243
Trevor, Jack and Tusia 76, 78, 79, 138, 231, 232
Trilobites (animal life) 109
Trinil, Javanese calotte 230
Trotter, Mildred ('Trot') 114, 116, 117, 135
Truth cannot contradict truth (Pope John Paul II) 155
tsetse fly, *Glossina morsitans* 23
Turner, William 113

UCLA Department of Anthropology and Sociology 106
Ullrich, Herbert 247
Umgazana River (cave deposit) 17
Unesco (United Nations Educational, Scientific and Cultural Organisation) 76, 242, 243, 248
Union of South Africa 8
United Party 184
University Education Bill, the Extension of 59, 189, 198
University
of Arizona 161
of Bophuthatswana 117
of California 100, 225
of Cape Town 14, 57
of Chicago 124, 138
of Cincinnati 213
of Edinburgh 174
of Fort Hare 57, 58
of London 167
of Michigan 119, 120, 184
of Mmabatho 117
of Natal 57
of North West 117
of Pennsylvania 141, 142, 243
of Queensland 212
of Rome 238, 239
of Sydney 212
of Uppsala 245
of the Witwatersrand *see* Wits
upright posture 28
upright stance and gait 27

Valentine, James 223
Vallentine, Isaac 19
Vallois, Henri V. 83, 86, 138
Van der Horst, Cornelius 23
Van der Merwe, Nic 127
Van Hoogstraten, Richard 160
Van Riet Lowe, C. 31, 46
Vesaliana, Room of (Yale University) 147
Vezère Valley (cave sites) 86
vestibular and proprioceptive sensory inputs 28
viscerotonia 208
Viking Fund 241
Medal in Physical Anthropology 135
von Linné, Carl (Carolus Linnaeus) 245

Washburn, Sherwood ('Sherry') 27, 124, 127, 138, 139, 141, 157, 219

Washington University 213
Washington University Medical School 111
Watt, John Mitchell 211
Webb, Revd Joe 31
Weidenreich, Franz 230
Wegner, Dietrich 248
Wells, Lawrence and Nicolette 1, 50, 51, 63, 70, 78, 174
Wenner-Gren Foundation for Anthropological Research 241
  see also Viking Fund
Werner-Reimers Foundation 241
Western Reserve University, Cleveland 132
When Smuts Goes (A. Keppel-Jones) 176
Williams, Eric 40
Wilson, J.T. 92, 212
Windhoek Museum 68
Wits 5, 10, 14, 23, 55, 60, 62, 69
  Dental School 179
  Department of Anatomy 23, 24, 33, 39, 40, 71, 101, 103, 104, 175, 179, 180, 183, 212, 237
  Department of Botany 23, 38
  Department of Pharmacology 211
  Faculty of Health Sciences 147
  Faculty of Medicine 22, 116, 123, 182, 191, 192, 211
  Kalahari Research Committee 122
  Medical Graduates' Association 214
  Old Medical School 25
  Surgery Department 180
  University Council 191
  Zoology Department 224
Witwatersrand Council of Education 237
Witwatersrand Medical Library 214
Wolfowitz, Brian 178
Woodburne, Russell T. 119
Wordsworth, William 20–21
World Archeological Congress (1986) 194
World Heritage List 248
World War II *see* Second World War

Yale University 145, 146
  Library of the History of Medicine 146, 147

Zambia, Kabwe cranium (Rhodesian Man) 81, 242
Zambia, Tonga people 23, 122
Zhoukoudian (Peking Man discovery site) 230
Zhoukoudian teeth 246
Zinjanthropus boisei 243
Zuckermann, Solly 76, 79, 88, 89, 177
Zwi, Saul 73
Zwigelaar, T. 244